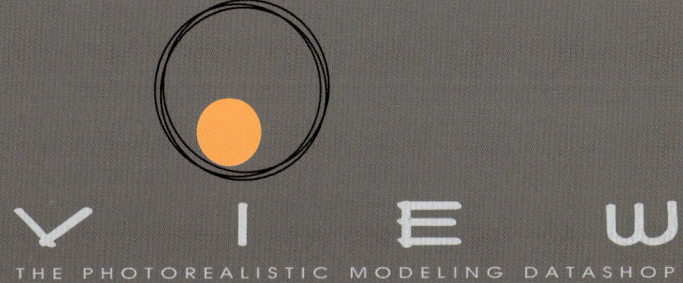

VIEW
THE PHOTOREALISTIC MODELING DATASHOP

| TABLE-WARE | VOL 1 |

VIEW
UL. PRZEDZALNIANA 8
15-688 BIALYSTOK, POLAND
NIP: 542-291-02-99

PHONES
PH: +48856745006
    +48858143039
FAX: +488711039

VIEW CREATES AND DEVELOPS PRODUCTS THAT HELP CG ARTISTS WITH THEIR 3D VISUALIZATIONS BY SPEEDING UP THEIR WORKFLOW AND INCREASING THEIR QUALITY OF OUTPUT. WHY WASTE COSTLY TIME IN DEVELOPMENT WHEN YOU CAN HAVE VIEW'S TAILOR ITSELF TO YOUR EXACT CG OUTPUT NEEDS?
SINCE 2008 WE ARE PROVIDING CG SERVICES FOR MANY SIGNIFICANT CLIENTS. BESIDES THIS WE'RE ALWAYS WILLING TO HELP CG FANS WITH THEIR JOURNEY THROUGH THE THIRD DIMENSIONAL WORLD AND THEREIN SHARE OUR VAST BODY OF KNOWLEDGE.

THE PHOTOREALISTIC MODELING DATASHOP

VIEW CREATES AND DEVELOPS PRODUCTS THAT HELP CG ARTISTS WITH THEIR 3D VISUALIZATIONS BY SPEEDING UP THEIR WORKFLOW AND INCREASING THEIR QUALITY OF OUTPUT. WHY WASTE COSTLY TIME IN DEVELOPMENT WHEN YOU CAN HAVE VIEW'S TAILOR ITSELF TO YOUR EXACT CG OUTPUT NEEDS?

ALL RIGHTS RESERVED - COPYING AND PUBLISHING WITHOUT PERMISSION IS FORBIDDEN

VIEW

UL. PRZEDZALNIANA 8
15-678 BIALYSTOK, POLAND
NIP: 552-291-02-99
PHONES
PH.: +48856745006
    +48858743039
FAX: +488711039

FREFACE

MADE WITH:

THE NEW DIGITAL AGE

ALL 3D MODELS ON OUR BOOK ARE CREATED
USING V-RAY 1.5 WITH 3DSMAX 2010 OR HIGHER.

IF HIGHER VERSION OF 3DSMAX HAS BEEN USED,
YOU WILL ALWAYS FIND 3DS, OBJ OR FBX FILE IN A ZIP ARCHIVE.
THESE FORMATS CAN BE OPENED USING ANY VERSION OF 3DSMAX.

MATERIALS AND TEXTURES ARE INCLUDED.
ALL MODELS USE QUALITY TEXTURES AND UNWRAPPED UVS.

ENJOY THE READING,
AND ALWAYS REMEMBER,

IF YOU WANT TO DO IT

RIGHT!

DO IT LIKE A PRO!

Δεν ελπίζω τίποτα.

Δε φοβάμαι τίποτα.

Είμαι λεύτερος

– NIKOS KAZANTZAKIS'S GRAVE

"I DON'T HOPE FOR ANYTHING, I DON'T FEAR ANYTHING, I'M FREE"

THE PHOTOREALISTIC MODELING DATASHOP

CONTENTS

1 11-14p

- FUNDAMENTAL KNOWLEDGE

1-1. OBSERVATION

1-2. PRACTICE

2 17-30p

- SYSTEM CONFIGURATION

2-1. WHAT IS V-RAY?

2-2. MONITOR CALIBRATION & SOFTWARE

2-3. LINEAR WORKFLOW

2-4. ABOUT THE BOOK
(THE METHOD WHICH USES BOOK)

TABLE-WARE VOL 1

MADE WITH:

THE NEW DIGITAL AGE

● ● ③ ●

ABOUT CATALOG   33 – 58p

● ● ● ④

– ABOUT SETTING

1. MODERN STYLE & CLASSICAL STYLE 50EA

61 – 260p

VIEW

THE PHOTOREALISTIC MODELING DATASHOP

● ○ ○ ○   11–14p

— FUNDAMENTAL KNOWLEDGE

1-1. OBSERVATION
1-2. PRACTICE

MADE WITH:
+ v·ray
THE NEW DIGITAL AGE

1. FUNDAMENTAL KNOWLEDGE

In order to be a professional 3D Artist you got to have that basic "Fundamental Knowledge" and to know how and where to use it, followed by logic explanations of your actions. In other words, to have good understanding of "How things working in our REAL world" cos' this is the real down fundamental.

That's it. Turn off your computer, go to sleep. End of story.

No really, that's sounds really clear but I'm pretty sure that not many of the professional 3D Artists can explain that either...
So let me just explain to you why it is important to have that "Down Fundamental" stuff. Well lets assume for a second that you learned how to play tennis and now you want to get in the tournament and play with the "big guys" but before you do that you want to practice before the big game. Well you take your tennis bet, get to the court, and start to practice, but suddenly things are getting really ruff! This is no longer your back yard, where you could hit the ball for hours up against the wall, this is more serious stuff going on here...you hear your trainer " Your foot is a mess, get back on your fundamental foot work"..."Your swing is a mess, you're not going from the high to low, you're going from high to high"...and then you realize that it's really harder than you thought, and you try to figure why?

Well...this is just about it, "If you don't have your fundamental down, you work twice harder to make that shot".

This is most common mistake, I've done it as well, and most of the beginners doing it.

"You running too fast!"

You make a tutorial, or get someone else scene and use it for your renders, you get so exited from the first results, that you think you're so awesome 3D artist, and you just go on with it, and forget about the basic learning and preparations that you need to do, before you dive into this temptation of using someone's settings or scene. And then what happened? you keep on making tutorials and try to build the whole "picture" from different pieces of knowledge that you find in internet, and you ask for other user's rendering settings..and you test them..

Well personally I didn't do it, but I see every day lots of users doing it. You think that other persons rendering settings going to help you becoming a greatest 3D Artist of all times? Well, I don't think so. Sorry but it's true.

Have you ever tried reading or understanding "rules of light" or "rules of design", and understand the basic fundamental of those things? Well, believe me, if you do that it will be much easier from you to read V-Ray manual and start practising your scene. Cos' you will know what you after, and you will know what to expect after "Render" ends.

The biggest enemy of your success is your laziness, be patient and it will come, some learn faster some slower, but at the end everybody getting it. Of course it is always good if there is someone to guide you, and to instruct you, that's why I'm here, telling you this, and if you are still reading it, it means that you want to become PRO, maybe the next big thing!

So this is all theoretical, "You got to learn the theory, before you start your car". OK I think you got it already, now how do you do it.

Quick basic rules and it will not cost you any money:

1. FUNDAMENTAL KNOWLEDGE

1. Observation – you have to learn to observe "life" correctly, and to picture with your mind all that your eyes see, remember "The best library you ever have, is the one that in your head". You must be able to pull out every scene from your memory and apply it to your work. Your mind library is the best tool you're ever have, and it will grow everyday as soon as you start filling it."So, how YOU do it?" Well, don't just look, " Observe", behold the nature greatness creativity, pay attention to all small details, try to remember every small part of what you see, try to feel the light and it's behaviour, pay good attention to the dark and bright parts, where the light hit the ground and creates the shadows.

If you're in the industrial zone, or just assume you're in the underground metro station, pay attention to the marble, to the reflections it creates. Is it sharp? or is it gets blurred as it fades away? From what angle can you see the lamp reflected in it? Try to understand that basic behaviour of materials. It doesn't matter if it's wood or stone, marble plate or wooden door, it's the feeling that you get while observing it. Practice your eye to get natural colours, behaviour of the materials and their interaction with our world, take more than few seconds to watch it, stare at it, go deep to the material.

2. Practice – the best way of practising the "Light" and understanding it's behaviour, is to paint it, even if you can't paint, do it! This is the most important procedure, when your eyes watching, transmitting it to the brain, and trying to apply it on the paper, is the best method of digging that fundamental stuff. If you have difficulties of seeing and applying of what you see, take a picture and print it, put it near by, and try to paint what you have pictured.
The more you practice the better you get!
Make it a rule number one, if you want to be good at CG fist create it on paper, it must not be perfect, but you have to give it at lest 2 weeks before you sit at your computer and try to create something.

Another good practice is your photo-camera, the better it is, the better practice you get. I've noticed that I could significantly improve my control over 3D "VrayPhysicalCam" while making tests with real reflex camera. I got Nikon D70, it's a bit old, but to get full understanding of f/number, shutter speed, ISO, Depth of Field...etc. You need to dig it first. This practice will be the most effective one you could ever get, plus it's fun, I've made it my hobby.

The bottom line of all this, if you read books, do tutorials, sit nights and days over computer trying to figure stuff out, will not bring you further as much as real time practice outside. In order to learn the fundamental you got to feel them. You got to learn first working with your hands, and that is the best training you can ever experience, believe me I know.

Well there is also lots of short cuts of different sources to do it, and if you want to invest some money it might save you lots of time: You can take a part in different practising with professional trainers, you can also join the community and get your work criticised, improving it step by step or just ask for advice, but, remember, every advice is something that comes from the personal experience of that specific person, and you might have just other luck experiencing that.

So if you have some spare money and you want to improve yourself, take painting lessons, or a photo course, do seminars and never stop learning, every time you try to do something, do it with all your heart, like it is the last thing you ever do.

If you want to build a perfect wall, every stone got to be perfectly aligned, with all your love and passion. Do it! It's now the time to take action, and enjoy your seeds that you grow after.

Remember the **power** is in your hands, and if you really, no mean **REALLY!** Want to do it **RIGHT**! Do it like a **PRO!**

VIEW THE SKY...

VIEW
THE PHOTOREALISTIC MODELING DATASHOP

- SYSTEM CONFIGURATION

2-1. WHAT IS V-RAY?

2-2. MONITOR CALIBRATION & SOFTWARE

2-3. LINEAR WORKFLOW

2-4. ABOUT THE BOOK
   (THE METHOD WHICH USES BOOK)

MADE WITH:

THE NEW DIGITAL AGE

2. SYSTEM CONFIGURATION

2-1. WHAT IS V-RAY?

Here is a good start, for getting to understand basic principles of your future working tools. Yes! You guessed it right I'm talking about V-Ray. My advice to you on this one will be, "If you want to be professional, have perspective!"

Know the basis, know to give basic explanation for those who asking, and know what will happened in the future development. If you can answer those questions, and back them up with evidence, then it's a good start for becoming a professional 3D Artist, in other words you'll have a perspective!

What is V-Ray?

V-Ray is a "Rendering Engine" (according to the Wikipedia) and was developed in Bulgaria by two guys, Vladimir Koylazov and Peter Mitev around 1997. Officially realised for beta testing in 2001 when they established a company named "Chaos Group".

Some of you might get confused in trying to explain those basic principles to someone new if the field, so here's more understandable explanation:

Rendering is a process, computer calculation to produce an "image" out of 3D program. Render=Image.

Engine is an application, a plug-in that can be installed inside the 3D program in order to produce this computer calculation.

In other words V-Ray is a plug-in that runs a mathematical algorithm computation on variety of 3D software so it can produce an "output", which is our image.

How doe's it work?

Just to make this short and simple. V-Ray is a "Ray Tracing" based engine, it shoots rays and track them back, rays that are being "traced" in the conjunction with 3D environment, such as 3D objects, lights, or even sky! These "traced rays" include colour information and are able to create a colour "dot", as known as "pixel". Dot after dot, rays are being traced, and render process being proceeded, and at the end of the it we get an "Image".

On which 3D software V-Ray can be installed?

Good news for you guys, cos' V-Ray can be installed on almost every 3D application today. Well, at least the big boys already using it, such as: 3Ds Max, Maya, Cinema 4D, Sketch Up, Rhino, Blender. As well as coming soon after beta testing V-Ray for Light wave and Soft Image XSI.
Isn't it just great? All that V-Ray love is spreading all around to all available 3D applications, and if you ask me it will be pretty much the same thing for all of the available 3D applications, but will manage the advantage of every 3D software individually, and what do I mean by that is...

Pretty nice don't you think?

Well, all the 3D programs have their basic Rendering options/Rendering Engines, and some of them are pretty good, but V-Ray has proved itself over the years as the fastest, easiest and one of the highest quality " Rendering Engine" on the market. The popularity of V-Ray is growing so fast every day that I'm pretty sure it will be the main "Rendering" software to be used in the entertainment industry future days to come. (See it for yourself – Architectural Visualization Industry Survey)
Well good! But I've heard that there are much better Rendering Engines, such as Mental Ray, Final Render, Brazil.
Heck yeah! And No! It is exactly what I was talking about in the introduction chapter, "it is not about what software you use, it's
about how you use it" (and what can you get out of it !). But let me just note something that you might be missing. With V-Ray knowledge, could be much easier to get a job plus it is the fastest rendering engine out there, and fast results matter! So if you are a freelancer conceder that Time=Money, that is the main key of V-Ray and by explaining that to your customer could position you at the right priority.
Wait there is more... V-Ray is about to go much more faster, so fast that you will not need to spend hours of sitting and waiting for your render to be done, I'm talking about minutes, may be even seconds, for one complicated image to be rendered. For those who didn't know Render is a process that depends on your computer quality hardware, more RAM, faster processor=faster "Rendering" times.

2. SYSTEM CONFIGURATION

2-2. MONITOR CALIBRATION & SOFTWARE

Monitor calibration – Is a process that brings monitor colours to it's origin, to the base manufacturer profile. Let's start from the beginning, it will help us to understand the process.

Monitors like every other technical equipment get warm while working, that is the main reason why monitors start to lose their colour productively, as we all know every monitor works on 3 colours RGB (red, green, blue) and the lost of these colours happens not uniformly on each colour channel. Some monitors lose more Red, some – Green, others- Blue...it can be even the same monitor model from the same manufacturer, but after a while each one of them will show the same image differently, because of that "colour loosing phenomenon".

"So what do we do ? How can we bring those colours back ?"

Here comes a handy one very nice equipment that brings all lost colours back and sets up our monitor profile to the correct one (which is gamma 2.2).
For "Monitor Calibration" I use "eye-one display 2", it's a grate calibrtion device and I'm personally very satisfied with it, I've tried to use "Spider", but gave it back and took the "eye-one display 2" instead. So I warmly recommend it, if you want to have the right colours, and knock them down with your images, calibrate your monitor. I know you might say...

"But most people are not using it, how come I should ?"

Well simply because Pro 3D Artist, which can really give you a good feedback on your image, or accept you to work with them, do calibrate their monitors, yes, I'm talking about the big boys, so if you want to play in their league, you better come with correct colours.

Software – OK so after you have calibrated your monitor, you can calibrate your 3Ds Max and set it to gamma 2.2, it doesn't have to be one after another, these are two separated procedures, but you should make at least this one right now!

Go to 3Ds Max program, at the main control panel, select "Customise – > Preferences – > Gamma and LUT" and make these changes.

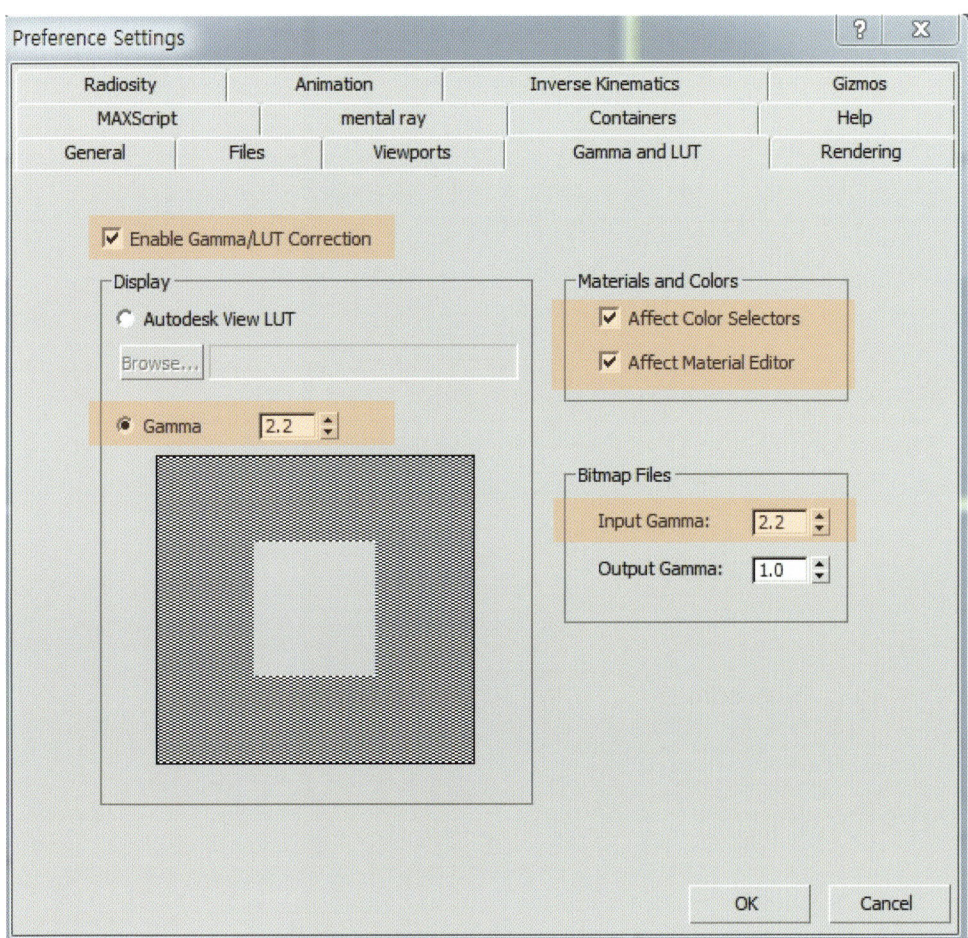

OK now we have set our software to the professional mode.

2. SYSTEM CONFIGURATION

2-2. MONITOR CALIBRATION & SOFTWARE

What did we do exactly?

- We "Enabled Gamma /LUT colour correction" in our 3Ds Max System.

- We made it possible to "Load files with that Status.." -If you'll be loading some old scene with different old gamma setup, you will get message, so ignore it, otherwise you get the whole different render. So it is advisable to start a new scene from scratch on that mode.

- "Display" – we made it possible to see all correctly it in our system, now 3Ds Max will display all the information with gamma 2.2 on our Monitor. By the way Photoshop got that option built in.

- "Affect Colour selection" and "Affect Material Editor" – Gamma 2.2 will take effect on our colours and materials appearance in Material Editor, so basically it's only the correct appearance.

- "Input Gamma" – all textures that will be imported in to 3Ds Max automatically will get gamma 2.2 colour correction. (Sometimes textures appear too dark, it is because of double gamma 2.2, what you can do is just change the gamma of the texture manually in the "open bitmap dialogue" while importing texture in to 3Ds Max.

congratulations you are on the right way!

You probably notice that your materials editor have become much brighter, and it doesn't feel right fyour eyes, well I got somgood news for you, it is the old colours management that was not right! Don't worry, after a week you'll get used to it, and you will be amazed to see how it's looks awful with gamma 1.0 setup. I went through this as well, so you get over it, don't worry.

Now look here, if you just click on the colour selector at your material you will notice how much colour range it got now. We missed all that colour, and you didn't even know about it! well now you got it.

OK great! Now if you could just explain what happened here and why are we doing this?

Well to define it better let me just give you a simple example on what is happening. Let's pretend that we are sitting in the dark room, and suddenly going outside, and it's really shiny day, well what happened? Yes, you are going temporary blind.

Imagine the opposite situation, when you're at some party, and you decide to go home, you step outside and suddenly some one closes the door, and you can not see anything, but not for long, if you keep standing there for about a minute your eyes will adopt to the environment so you can see something.

Let's just analyse it, I'm not a doctor but I assume that we have some sort of sensor in our eyes that adjust light intensity so it would be good enough to see the details in extremely different situations. If it is very bright environment we still can see that shadowed places and not to get overexposed on the white colour, what probably would happen if we took a picture with reflex camera. Vice versa in a dark environment we still be able to see something, even though light is very low, and I really doubt that we could take a picture in such situation.

Anyway the logic behind this is simple, this "eye sensor" allows the light to be less intensive in the dark/bright situations and that allows us to see more details, and not to over-bright or over-dark image that our brain creates, that is what we want to achieve in 3D.

So what happens in CG, we are not going out side and have to adopt " our selves" to the environment, we stay in one place and this is our starting point, gamma 2.2 (monitor and software calibration) at this starting point we are trying to adopt the "3D" to it's origin, and the name of this starting point is:

2. SYSTEM CONFIGURATION

2-3. LINEAR WORKFLOW

What's it all about?

To start, I will simply show an example. Also note that this tutorials especially applies when you use GI in your scene.

The image on the right is an interior, with one very large opening. As you can see, there is bright light coming in, resulting in blown out areas around the opening, and very dark regions closer to the camera.
I would like to brighten the image up, so I increase the light strength. Result is even more blown out areas, and still not that bright in the foreground.

You would expect an interior with such a big opening to let the light travel much deeper into the scene...

Gamma 2.2 - wow!

The image on the right is the same as the first one, but I applied a gamma 2.2 curve to it (for example in photoshop you can easily change the gamma of an image). After this conversion, note that there is in fact much more visible detail in the image than you would think at first glance.

After this photoshop gamma correction, the image does look a bit flat and washed out. This is because I didn't change anything to the materials in the scene. The material in the gamma corrected image is in fact much brighter compared to the material used in the first image (the material is also gamma corrected, so it gets brighter).

So I adjust the material to be a similar grey for the gamma 2.2 version, render again and apply the gamma 2.2 curve again. That's already better, and now the lighting looks pretty decent, much more what you would expect in an interior scene with such a big opening and only bright materials

So what's happening?

The base problem is simple. Vray and 3ds Max do their calculations in 'linear space'. The vray camera has a 'linear' response to light. This means that it is working in gamma 1.0 space. All this is no problem, but since max and vray are setup to work in gamma 1.0 space, it also assumes that you view the output on a device that has a linear curve or gamma 1.0...
However your computer monitor is not gamma 1.0 at all, in fact most CRT monitors are by default gamma 2.5! Since we didn't tell max and vray about this, our images are actually displayed much darker!!!
Max does have the tools to specify what gamma our monitor has, so it can counteract for this and display our images correctly - in other words 'brighter'.

Old habits die hard...

When this gamma 2.2 setup first popped up, many people (including me) didn't see the point of using it. This is because we were so used to applying all kinds of tricks and fakes to our scenes to make sure the output looked ok.

For example with interior scenes. There was always a problem that light didn't travel far enough into the scene. Increasing the environment light resulted in very burnt out areas around the window openings. To brighten things up, invisible lights were placed in the rooms, or things like exponential color mapping was invented to remove the burnt out areas.

There was also always an extreme color bleed. If you put a red table in the room, the lighting in the room tinted red because of the strong color bleed. So we started using desaturation on the GI solution, over-ride the materials with white and use saved IR maps with the red color back in place, etc... we had to trust in all kinds of fakes to make it look real...
Another common problem were hdri maps. On the right are two examples of the same scene lit by the same hdri map. The first one is the old habit, the second one is with my gamma 2.2 setup. The last image is how the original hdri map looks like. I think it's clear that the second render looks a lot more like the lighting in the original hdri map, compared to the oversaturated, high contrast first image. Also note the reflection in the sphere in the first render, it doesn't look like the hdri map at all (way more contrast!)

2. SYSTEM CONFIGURATION

2-3. LINEAR WORKFLOW

First 'problem' - choosing colors

I make the color of the walls a dark saturated green. Render the image: it doesn't look so dark and saturated anymore... This is because we are actually choosing colors in 1.0 space, and we are rendering in gamma 2.2 space...

The solution is easy, in the gamma & LUT preferences, just select both the "**affect color selectors and material editor**" options. When you press ok, you will see the interface changing drastically!

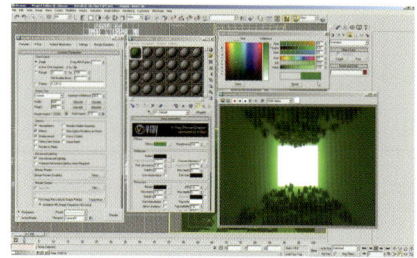

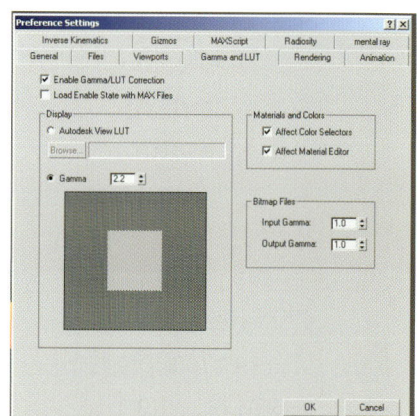

Look at the last image from this step. Both **material editor and color selectors are now gamma 2.2 corrected**. This might look very strange at first, but you'll get used to it. For me, when I revert back to the original look, it all seems extremely dark... So it's just a habit!

This is an important step however, because otherwise you will be choosing colors and they will always look brighter and more washed out in your renderings! You can see how the green has changed in the color selector. This is closer to the green in the rendering than it was in the previous step (light and unsaturated)

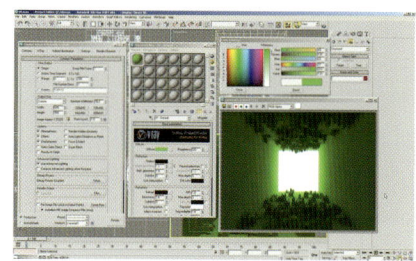

Second problem - washed out textures

Instead of a green color, I now add a texture map to the diffuse slot. I used this brick texture, which is very saturated. But in the rendering, it's again desaturated and washed out (compared to the original texture).

All your normal textures are in fact already gamma corrected. For example pictures from a digital camera usually have an srgb profile in them, to make sure they are displayed correctly on your gamma 2.2 monitor. So when you use this texture in max, without telling it is already gamma corrected, max will think it is a gamma 1.0 texture and it applies gamma 2.2 correction to it. But the texture already is gamma 2.2 corrected! **Applying gamma correction twice results in much brighter and desaturated colors...**

So we need to tell max this texture is already gamma 2.2 corrected. In the map's properties, click on the path button. Don't choose another texture, simply change the radio button from **'use system default gamma' to 'override' and fill in '2.2'**. Click open and notice how the textured material editor preview is already darker now, and also the texture in the viewport. This looks more like the original texture!

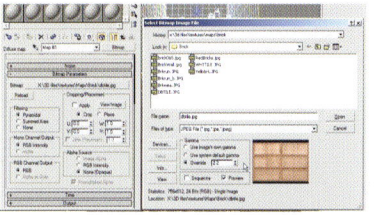

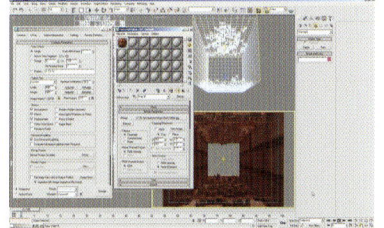

Render again, and the result is indeed more saturated like you would expect when using this texture.

Now you have two options for dealing with textures:
 - each time when you load a texture, specify its gamma in the bitmap loader
 - in the gamma & lut preferences, set **input gamma to 2.2**, so all your textures will be loaded into Max as gamma 2.2 textures

2. SYSTEM CONFIGURATION

2-3. LINEAR WORKFLOW

Third problem - saving the image

So far so good. I have a nice render, so I save it. Now I open it in an image viewer and suddenly it looks like crap: very dark and oversaturated.

The answer is again pretty simple. In max, we changed to gamma 2.2, which means everything is **'displayed' with a 2.2 correction**. The actual rendering doesn't have this correction 'baked in' however, it is only **previewed with gamma 2.2 in the max frame buffer.**

Again we have **several choices**:
- when saving the render, specify it should have a gamma 2.2 curve baked in
- adjust the gamma in photoshop
- leave it at 1.0, and adjust photoshop to work in linear space (1.0)

Advanced users who need to do very accurate post corrections and compositing, prefer to leave the image as is (1.0), and always use preview preferences to gamma correct the image. So they leave the image with gamma 1.0, or 'linear'. If you work with your images in this way, you are working in 'linear space'. However this isn't that easy, since **you have to setup your other applications to work in linear space** also to preview your image correctly.

Since most of the people don't need this, it is easier to '**burn in' the gamma 2.2 correction**. For example saving from max with the gamma 2.2 option or changing the gamma in photoshop both **burns in the gamma 2.2 curve into the image**

The vray frame buffer

Untill now, we always used the max VFB. However Vray has it's own VFB, which is in fact a lot better. To use it, go to the **Vray frame buffe**r and enable it. Next we have to **disable max's frame buffer**: in the common parameters tab, turn of the 'rendered frame window'.

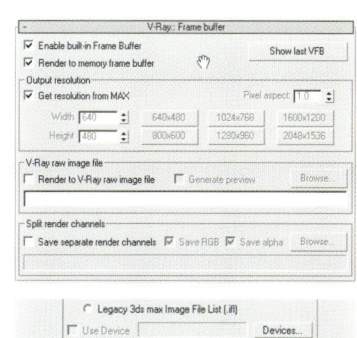

Now when I render again, the image in the Vray frame buffer looks also very dark, like the saved image when viewed in an image viewer outside max...
This is because the 3ds max gamma preferences DON'T affect the Vray VFB.

Luckily, the **Vray VFB has it's own gamma correction**. It's the small button labeled '**sRG**B' on the bottom of the frame buffer.

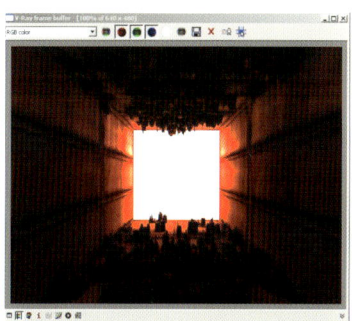

When saving the image, you have to set the gamma to 2.2 again to burn it in, because the **sRGB button is also only a display preview gamma correction**.

Color mapping

Color mapping provides a final way of **burning in the gamma 2.2 curve**. First we **disable the srgb button** again (the image will be very dark again).

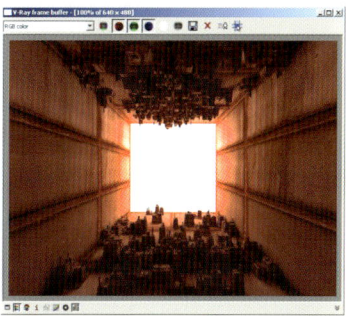

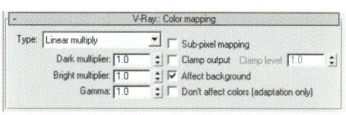

Now in the Vray tab, open the '**color mapping' rollout**. You can leave everything as is, you only have to change the **gamma to 2.2**.

Render the image again, and you'll see that it looks good now, without having to press the srgb button.

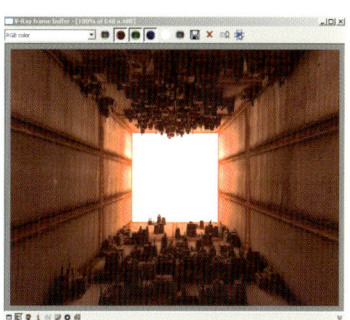

The big difference of this technique is that it applies the gamma 2.2 curve while rendering. The previous methods were all done after the rendering has finished (or as a preview only). **This difference is pretty important**, as it has serious consequences on image quality!

2. SYSTEM CONFIGURATION

2-4. ABOUT THE BOOK (THE METHOD WHICH USES THIS BOOK)

Merge
Insert objects from external 3ds max file into the current scene.

Just drag and drop.

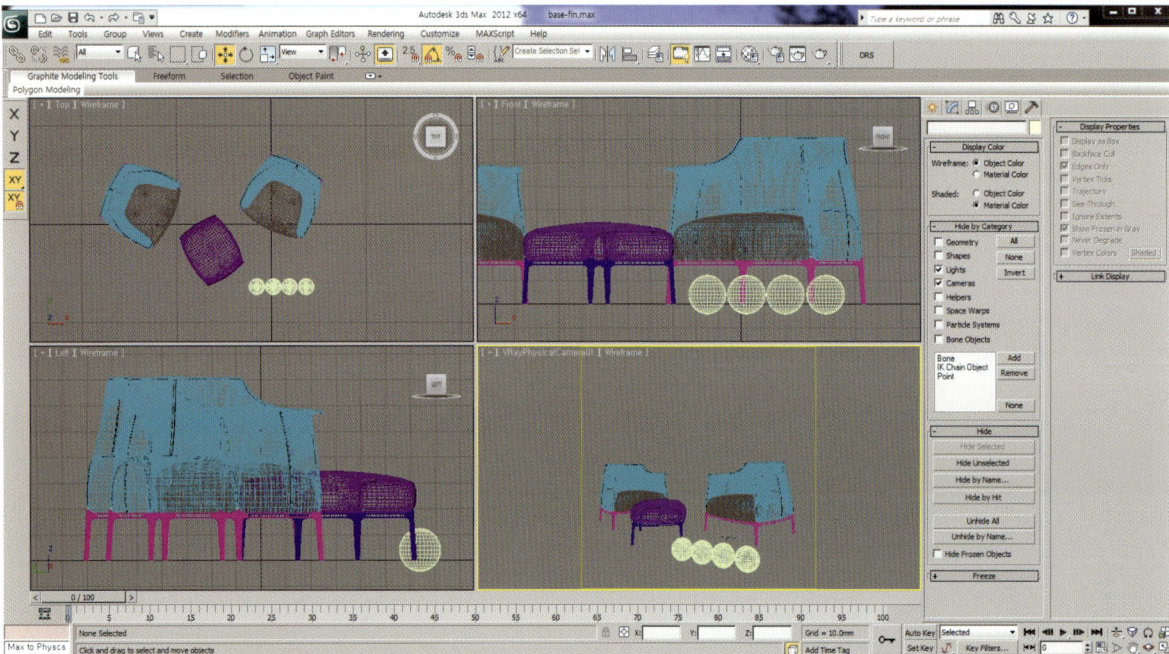

After when objects merge or drag and drop,
it render the scene.
Rendering image appears by right image fig1-1.

Like the obejects it is visible as not material.

It is because of the path.

FIG 1-1

Solution method #1

Command panel / utilities / Bitmap/photometric paths and Edit resources click.

1. Edit resources

2. Select Missing file

3. New path

4. Set path

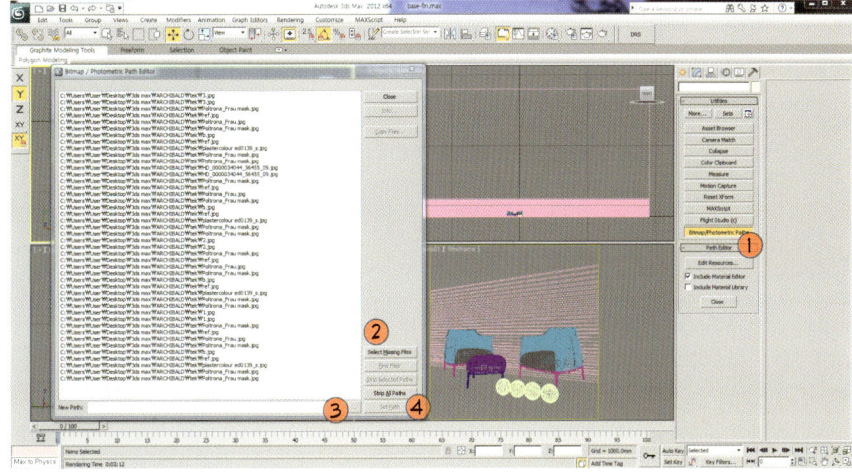

Solution method #2

Main toolbar / File / Manager / Asset Tracking click.

1. Bitmap Performance

2. Enable bitmap paging

3. New path

4. Set path

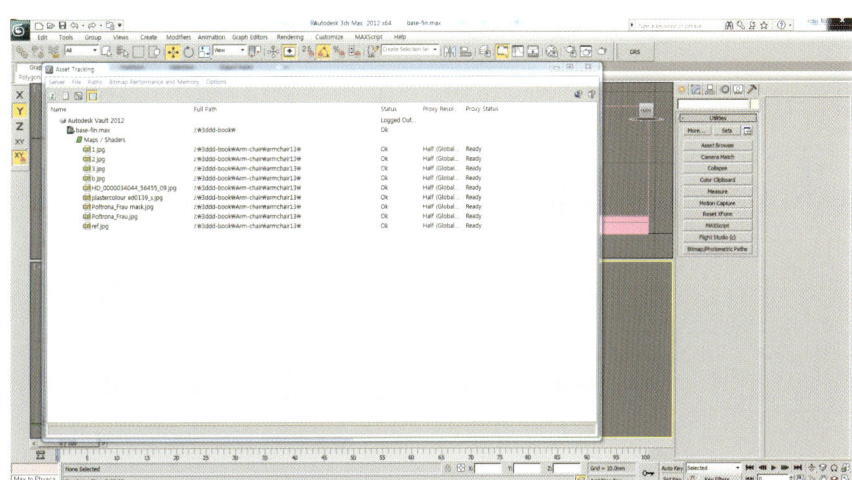

Rendering image!

FIG 1-2

TABLE-WARE VOL 1

MADE WITH:
 + v·ray
THE NEW DIGITAL AGE

○ ○ ③ ○

ABOUT CATALOG   33 — 58p

 TABLE-WARE | VOL 1

1

TABLE-WARE VOL. 1 INCLUDES 50 HIGHLY DETAILED 3D MODELS FOR ARCHITECTURAL VISUALIZATIONS. ALL OBJECTS ARE READY TO USE WITH TEXTURES AND SHADERS.

2

- FORMAT

MAX
3DS
OBJ - SIMPLE OBJECT WITHOUT TEXTURES AND MATERIALS (WITH MAPPING INCLUDED)
FBX - SIMPLE OBJECT WITHOUT MATERIALS (WITH MAPPING AND TEXTURES INCLUDED)

*.MAX - V-RAY 1.5 - OR HIGHER - WITH TEXTURES AND SHADERS
V-RAY - OBJECT PREPARED FOR V-RAY RENDERER (WITH TEXTURES AND SHADERS)

*. MAX 2010 - OR HIGHER

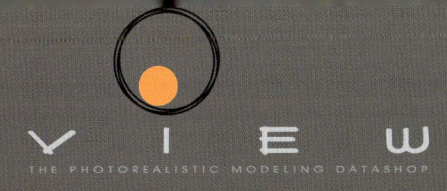

ЧАЙНЫЙ НАБОР   КОФЕЙНЫЙ НАБОР

1. 1.TABLE_WARE 001 - P61

2. 2.TABLE_WARE 002 - P65

TABLE-WARE VOL 1 MADE WITH 3DS MAX AND V-RAY

TABLE-WARE

1 3.TABLE_WARE 003 - P69

2 4.TABLE_WARE 004 - P73

1 5.TABLE_WARE 005 - P77

2 6.TABLE_WARE 006 - P81

TABLE-WARE

1 7.TABLE_WARE 007 - P85

2 8.TABLE_WARE 008 - P89

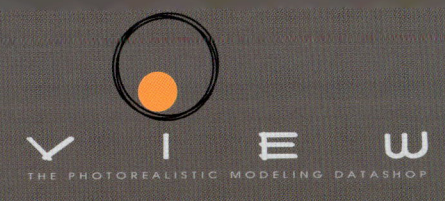 

1 9.TABLE_WARE 009 - P93

 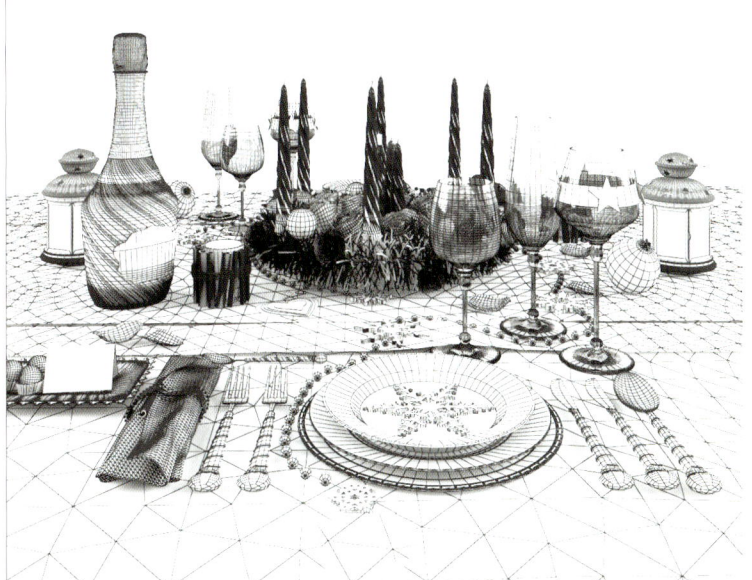

2 10.TABLE_WARE 010 - P97

TABLE-WARE VOL 1 MADE WITH 3DS MAX AND V-RAY

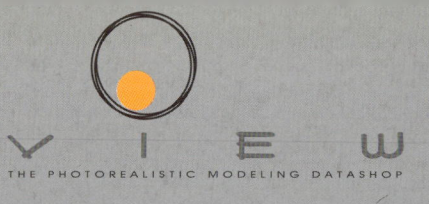

TABLE-WARE

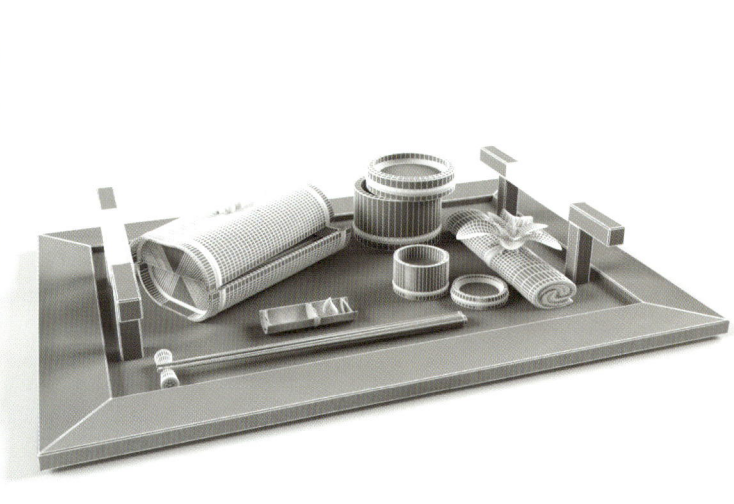

1 11.TABLE_WARE 011 - P101

2 12.TABLE_WARE 012 - P105

1 13. TABLE_WARE 013 - P109

2 14. TABLE_WARE 014 - P113

TABLE-WARE VOL I MADE WITH 3DS MAX AND V-RAY

TABLE-WARE

1 15.TABLE_WARE 011 - P117

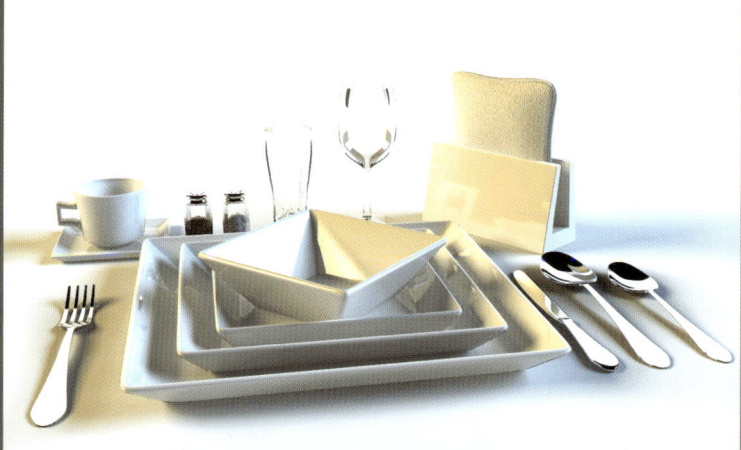

2 16.TABLE_WARE 016 - P121

1 17.TABLE_WARE 013 - P125

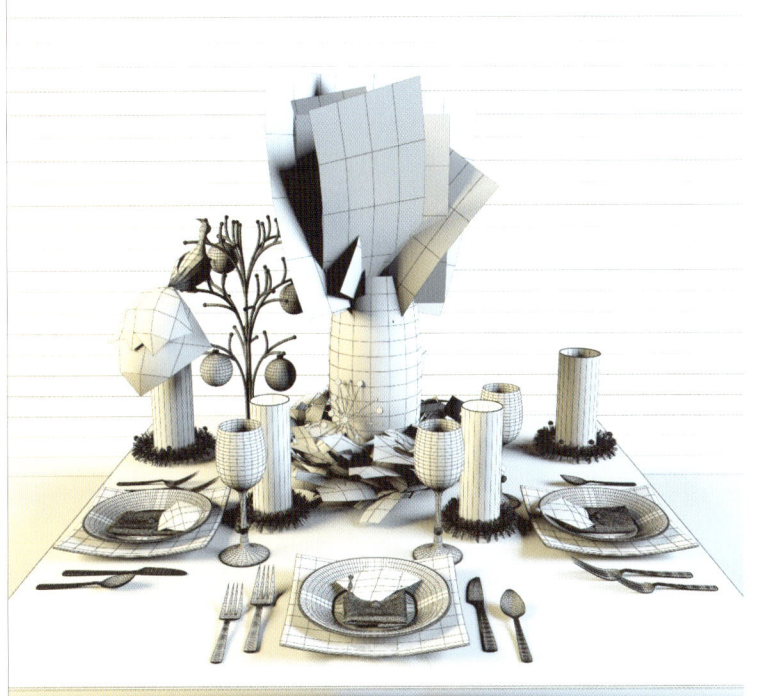

2 18.TABLE_WARE 018 - P129

TABLE-WARE VOL 1 MADE WITH 3DS MAX AND V-RAY

TABLE-WARE

1 19.TABLE_WARE 019 - P133

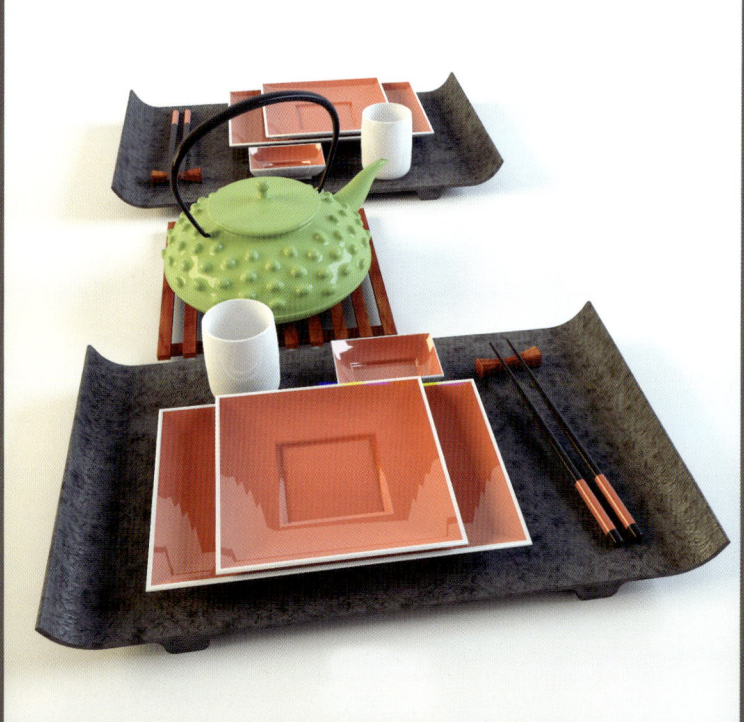

2 20.TABLE_WARE 020 - P137

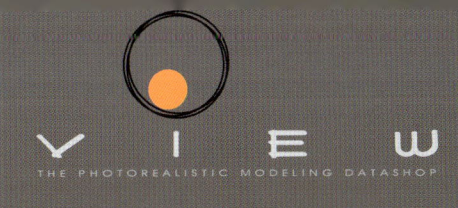

1 21.TABLE_WARE 021 - P141

2 22.TABLE_WARE 022 - P145

TABLE-WARE VOL 1 MADE WITH 3DS MAX AND V-RAY

TABLE-WARE

① 23.TABLE_WARE 023 - P149

② 24.TABLE_WARE 024 - P153

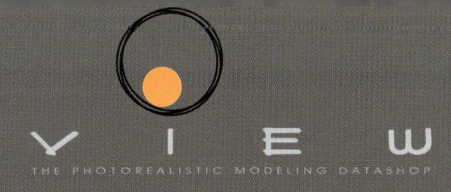

❶ 25.TABLE_WARE 025 - P157

❷ 26.TABLE_WARE 026 - P161

TABLE-WARE VOL 1 MADE WITH 3DS MAX AND V-RAY

TABLE-WARE

1 27.TABLE_WARE 027 - P165

2 28.TABLE_WARE 028 - P169

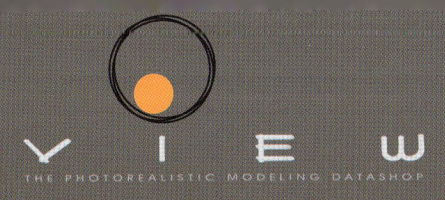

1 29. TABLE_WARE 029 - P173

2 30. TABLE_WARE 030 - P177

TABLE-WARE VOL 1 MADE WITH 3DS MAX AND V-RAY

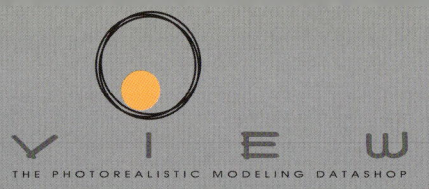

TABLE-WARE

① 31. TABLE_WARE 031 - P181

② 32. TABLE_WARE 032 - P185

1 33. TABLE_WARE 033 – P189

2 34. TABLE_WARE 034 – P193

TABLE-WARE VOL 1 MADE WITH 3DS MAX AND V-RAY

TABLE-WARE

1 35.TABLE_WARE 035 - P197

2 36.TABLE_WARE 036 - P201

1 37.TABLE_WARE 037 - P205

2 38.TABLE_WARE 038 - P209

TABLE-WARE VOL 1 MADE WITH 3DS MAX AND V-RAY

TABLE-WARE

1 39.TABLE_WARE 039 - P213

2 40.TABLE_WARE 040 - P217

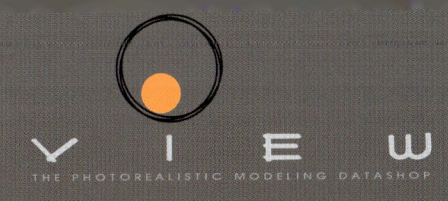

1 41. TABLE_WARE 041 - P221

2 42. TABLE_WARE 042 - P225

TABLE-WARE VOL 1 MADE WITH 3DS MAX AND V-RAY

TABLE-WARE

1 43.TABLE_WARE 043 - P229

2 44.TABLE_WARE 044 - P233

1 45.TABLE_WARE 045 – P237

2 46.TABLE_WARE 046 – P241

TABLE-WARE VOL 1 MADE WITH 3DS MAX AND V-RAY

TABLE-WARE

1. 47.TABLE_WARE 047 - P245

2. 48.TABLE_WARE 048 - P249

① 49. TABLE_WARE 049 - P253

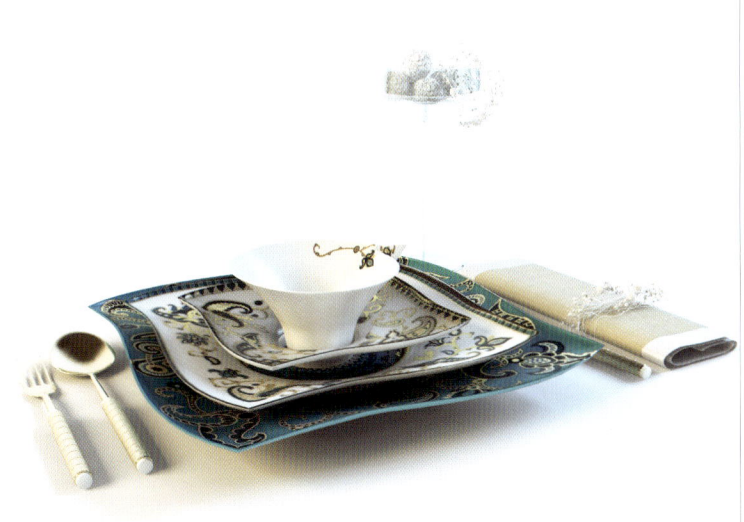

② 50. TABLE_WARE 050 - P257

TABLE-WARE VOL 1 MADE WITH 3DS MAX AND V-RAY

● ● ● ④

– ABOUT SETTING

1. MODERN STYLE &
   CLASSICAL STYLE 50EA

   61 – 260p

MADE WITH:

TABLE-WARE 001

3DS MAX 2010 + FBX (V-RAY)   19,308 KB   MODERN

MODIFIERS STACK
- NOT COLLAPSED
- COLLAPSED

TEXTURE
- NOT INCLUDED
- INCLUDED

MAPPING
- UNWRAP
- UVW MAP

TOTAL
- POLYS : 224,354
- VERTS : 158,627

TABLE_WARE

ЧАЙНЫЙ НАБОР

КОФЕЙНЫЙ НАБОР

MAX

3DS

FBX — SIMPLE OBJECT WITHOUT MATERIALS (WITH MAPPING AND TEXTURES INCLUDED)

*.MAX — V-RAY 1.5 — OR HIGHER — WITH TEXTURES AND SHADERS

V-RAY — OBJECT PREPARED FOR V-RAY RENDERER (WITH TEXTURES AND SHADERS)

*. MAX 2010 — OR HIGHER

ELEMENT OF MODEL

PARTIAL GROUP

1. TABLE_WARE-A
2. TABLE_WARE-B
3. TABLE_WARE-C
4. TABLE_WARE-D

TABLE-WARE VOL 1 MADE WITH 3DS MAX AND V-RAY

SLATE MATERIAL EDITOR

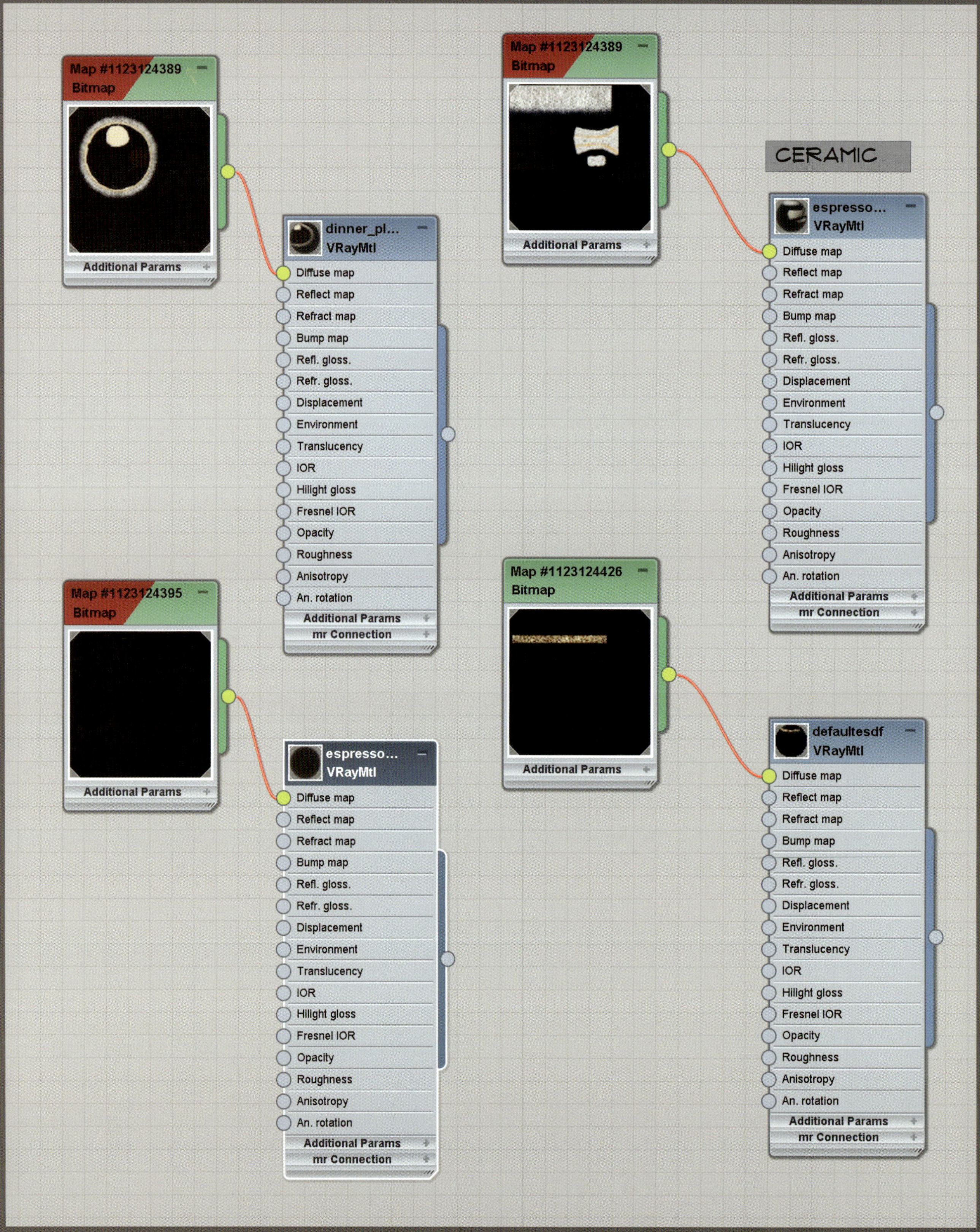

CERAMIC

TABLE-WARE VOL I MADE WITH 3DS MAX AND V-RAY

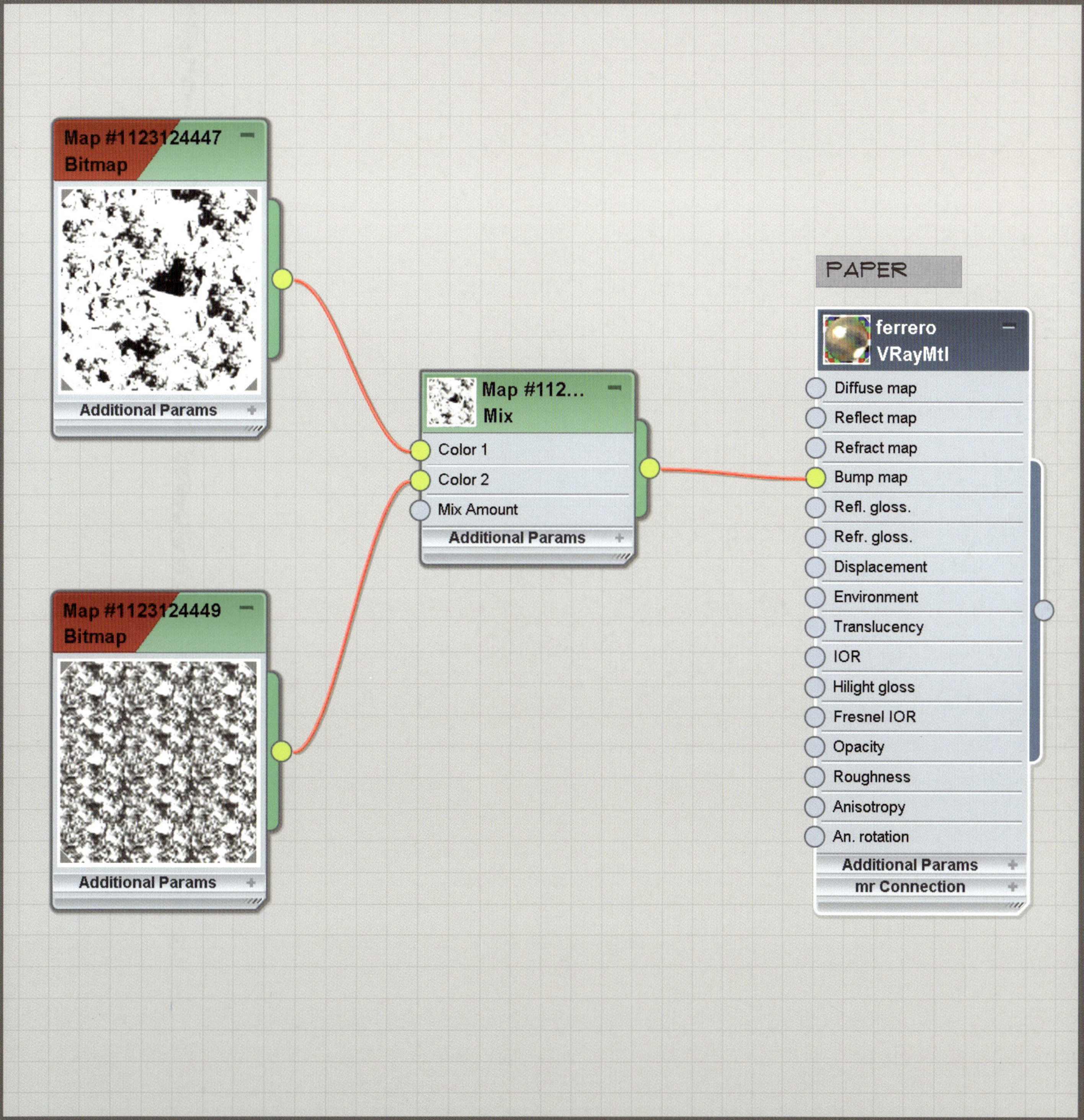

TABLE-WARE 002

3DS MAX 2010 + FBX (V-RAY) | 5,680 KB | MODERN

MODIFIERS STACK
- NOT COLLAPSED
- COLLAPSED

TEXTURE
- NOT INCLUDED
- INCLUDED

MAPPING
- UNWRAP
- UVW MAP

TOTAL
- POLYS : 193,604
- VERTS : 112,174

TABLE_WARE

MAX
3DS
FBX - SIMPLE OBJECT WITHOUT MATERIALS (WITH MAPPING AND TEXTURES INCLUDED)
*.MAX - V-RAY 1.5 - OR HIGHER - WITH TEXTURES AND SHADERS
V-RAY - OBJECT PREPARED FOR V-RAY RENDERER (WITH TEXTURES AND SHADERS)

*. MAX 2010 - OR HIGHER

ELEMENT OF MODEL

PARTIAL GROUP

1. TABLE_WARE-A 2. TABLE_WARE-B 3. TABLE_WARE-C 4. TABLE_WARE-D

TABLE-WARE VOL 1 MADE WITH 3DS MAX AND V-RAY

SLATE MATERIAL EDITOR

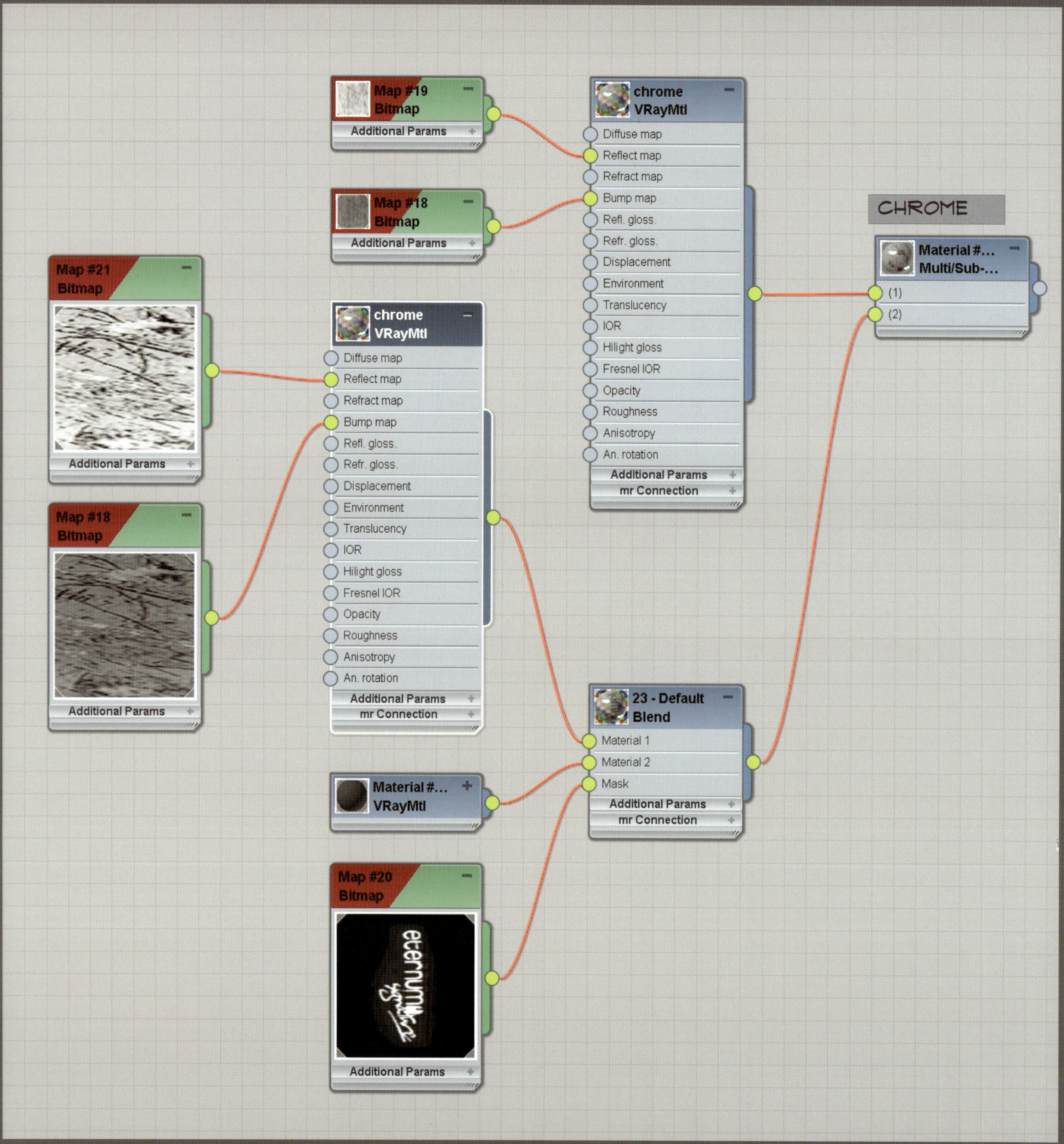

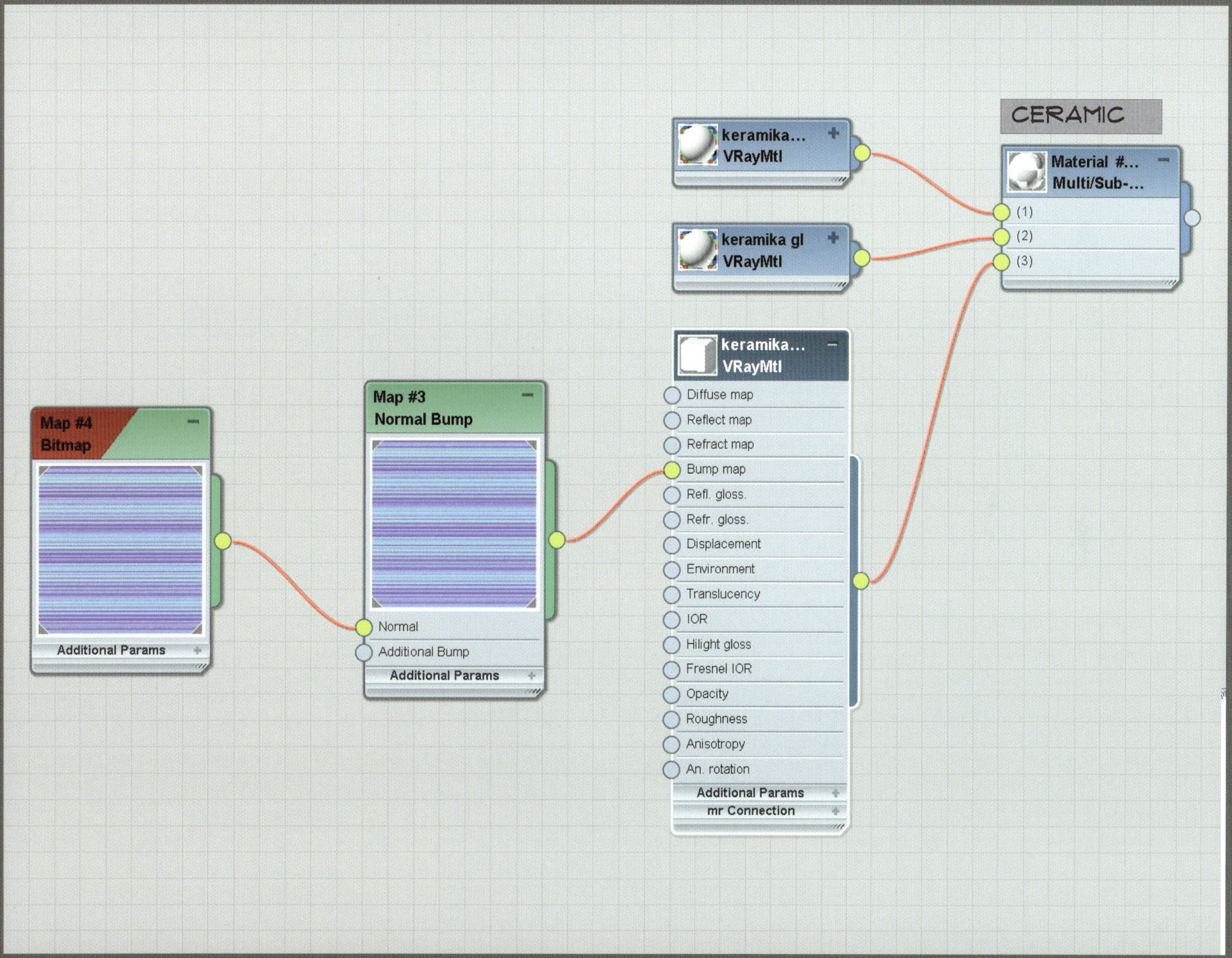

TABLE-WARE 003

MODIFIERS STACK
NOT COLLAPSED
COLLAPSED

TEXTURE
NOT INCLUDED
INCLUDED

MAPPING
UNWRAP
UVW MAP

TOTAL
POLYS : 241,532
VERTS : 158,336

 3DS MAX 2010 + FBX (V-RAY)    10,400 KB   ☆ MODERN

TABLE_WARE

MAX

3DS

FBX – SIMPLE OBJECT WITHOUT MATERIALS (WITH MAPPING AND TEXTURES INCLUDED)

*.MAX – V-RAY 1.5 – OR HIGHER – WITH TEXTURES AND SHADERS

V-RAY – OBJECT PREPARED FOR V-RAY RENDERER (WITH TEXTURES AND SHADERS)

*. MAX 2010 – OR HIGHER

ELEMENT OF MODEL

PARTIAL GROUP

1. TABLE_WARE-A   2. TABLE_WARE-B   3. TABLE_WARE-C   4. TABLE_WARE-D

TABLE-WARE VOL I MADE WITH 3DS MAX AND V-RAY

SLATE MATERIAL EDITOR

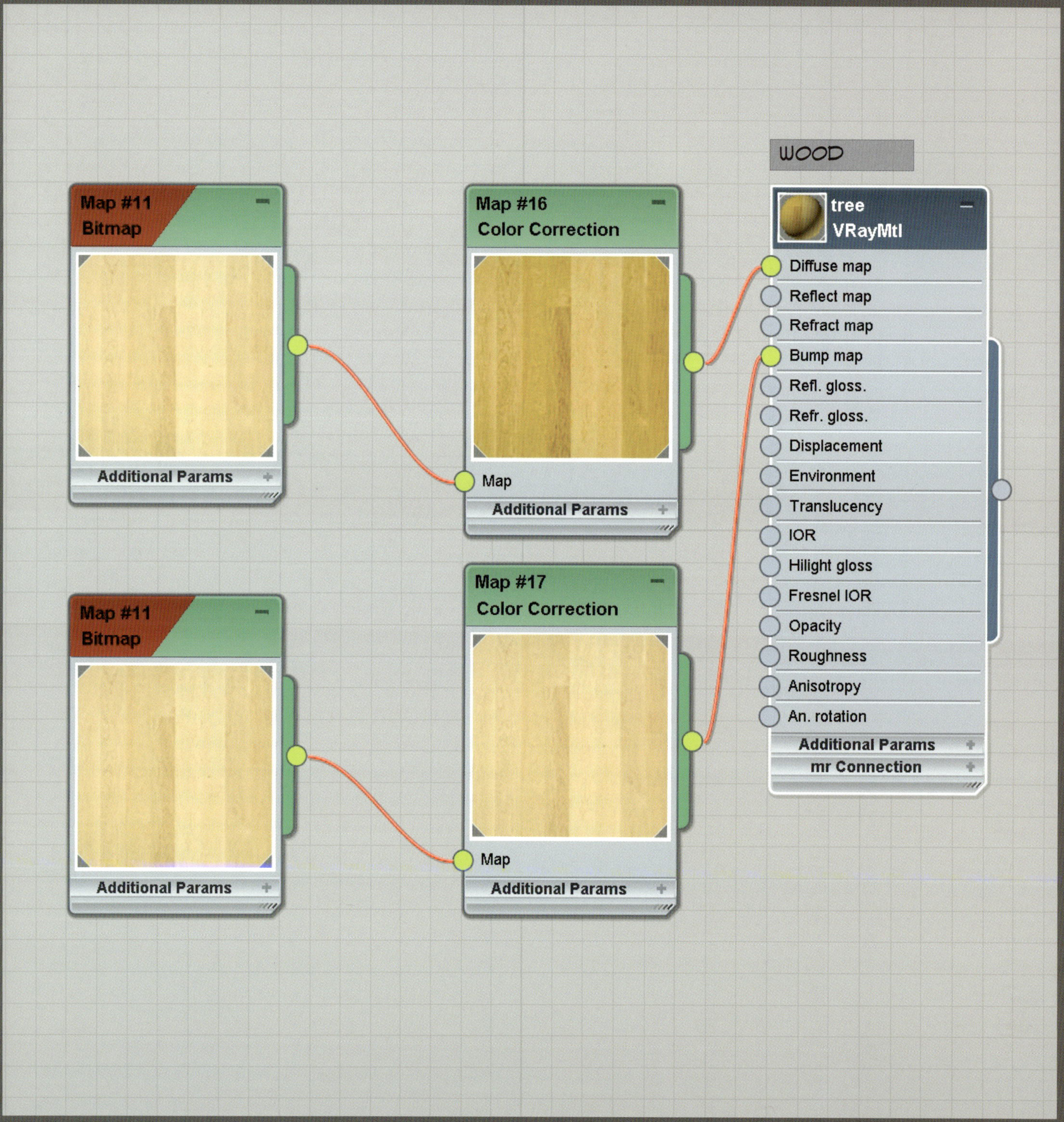

SLATE MATERIAL EDITOR

TABLE-WARE VOL 1 MADE WITH 3DS MAX AND V-RAY

TABLE-WARE 004

3DS MAX 2010 + FBX (V-RAY)   36,608 KB   MODERN

MODIFIERS STACK
NOT COLLAPSED
COLLAPSED

TEXTURE
NOT INCLUDED
INCLUDED

MAPPING
UNWRAP
UVW MAP

TOTAL
POLYS : 261,084
VERTS : 261,244

TABLE_WARE

MAX
3DS
FBX - SIMPLE OBJECT WITHOUT MATERIALS (WITH MAPPING AND TEXTURES INCLUDED)
*.MAX - V-RAY 1.5 - OR HIGHER - WITH TEXTURES AND SHADERS
V-RAY - OBJECT PREPARED FOR V-RAY RENDERER (WITH TEXTURES AND SHADERS)

*. MAX 2010 - OR HIGHER

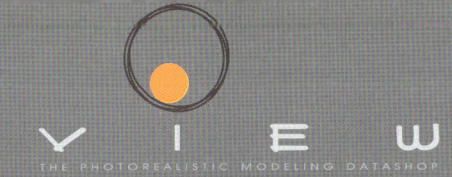

ELEMENT OF MODEL

PARTIAL GROUP

1. TABLE_WARE-A   2. TABLE_WARE-B   3. TABLE_WARE-C   4. TABLE_WARE-D

TABLE-WARE VOL I MADE WITH 3DS MAX AND V-RAY

SLATE MATERIAL EDITOR

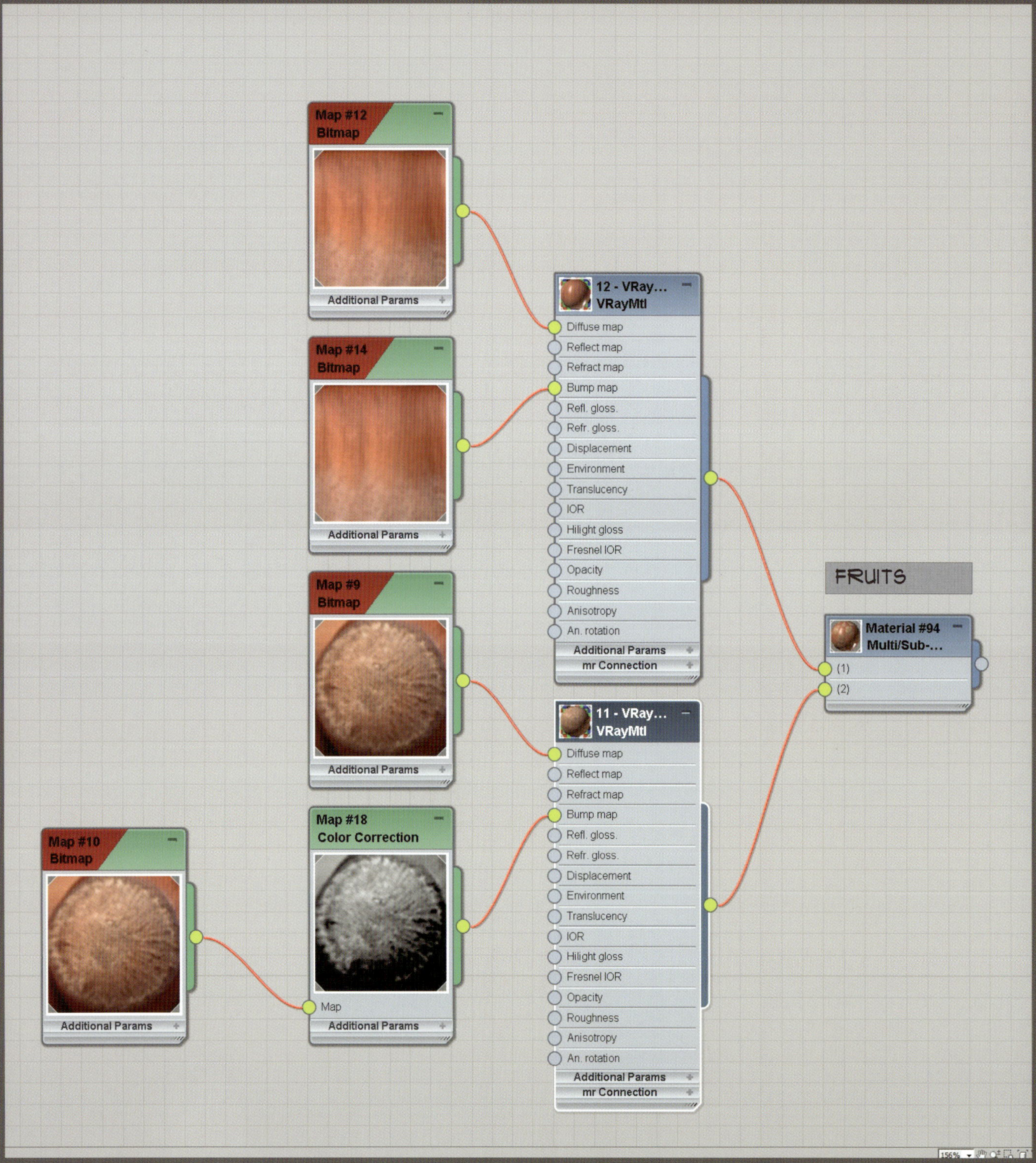

SLATE MATERIAL EDITOR

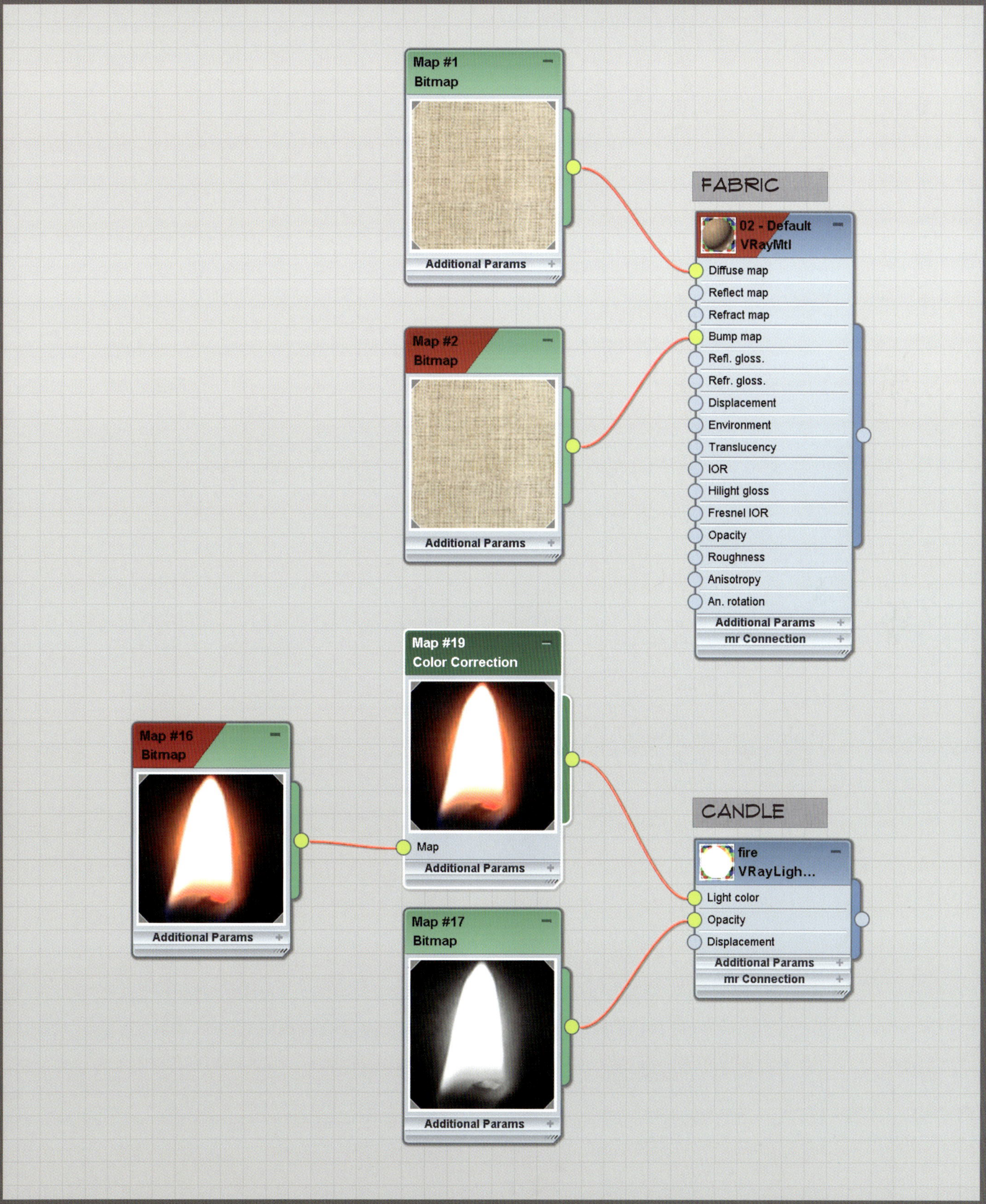

TABLE-WARE VOL I MADE WITH 3DS MAX AND V-RAY

TABLE-WARE 005

✱ 3DS MAX 2010 + FBX (V-RAY)   ◯ 6,388 KB   ☆ CLASSICAL

MODIFIERS STACK
- NOT COLLAPSED
- COLLAPSED

TEXTURE
- NOT INCLUDED
- INCLUDED

MAPPING
- UNWRAP
- UVW MAP

TOTAL
- POLYS : 216,110
- VERTS : 226,131

TABLE_WARE

MAX

3DS

FBX – SIMPLE OBJECT WITHOUT MATERIALS (WITH MAPPING AND TEXTURES INCLUDED)

*.MAX – V-RAY 1.5 – OR HIGHER – WITH TEXTURES AND SHADERS

V-RAY – OBJECT PREPARED FOR V-RAY RENDERER (WITH TEXTURES AND SHADERS)

*. MAX 2010 – OR HIGHER

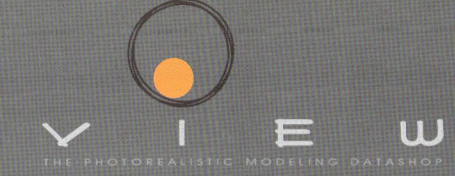

ELEMENT OF MODEL

PARTIAL GROUP

1. TABLE_WARE-A   2. TABLE_WARE-B   3. TABLE_WARE-C   4. TABLE_WARE-D

TABLE-WARE VOL 1 MADE WITH 3DS MAX AND V-RAY

SLATE MATERIAL EDITOR

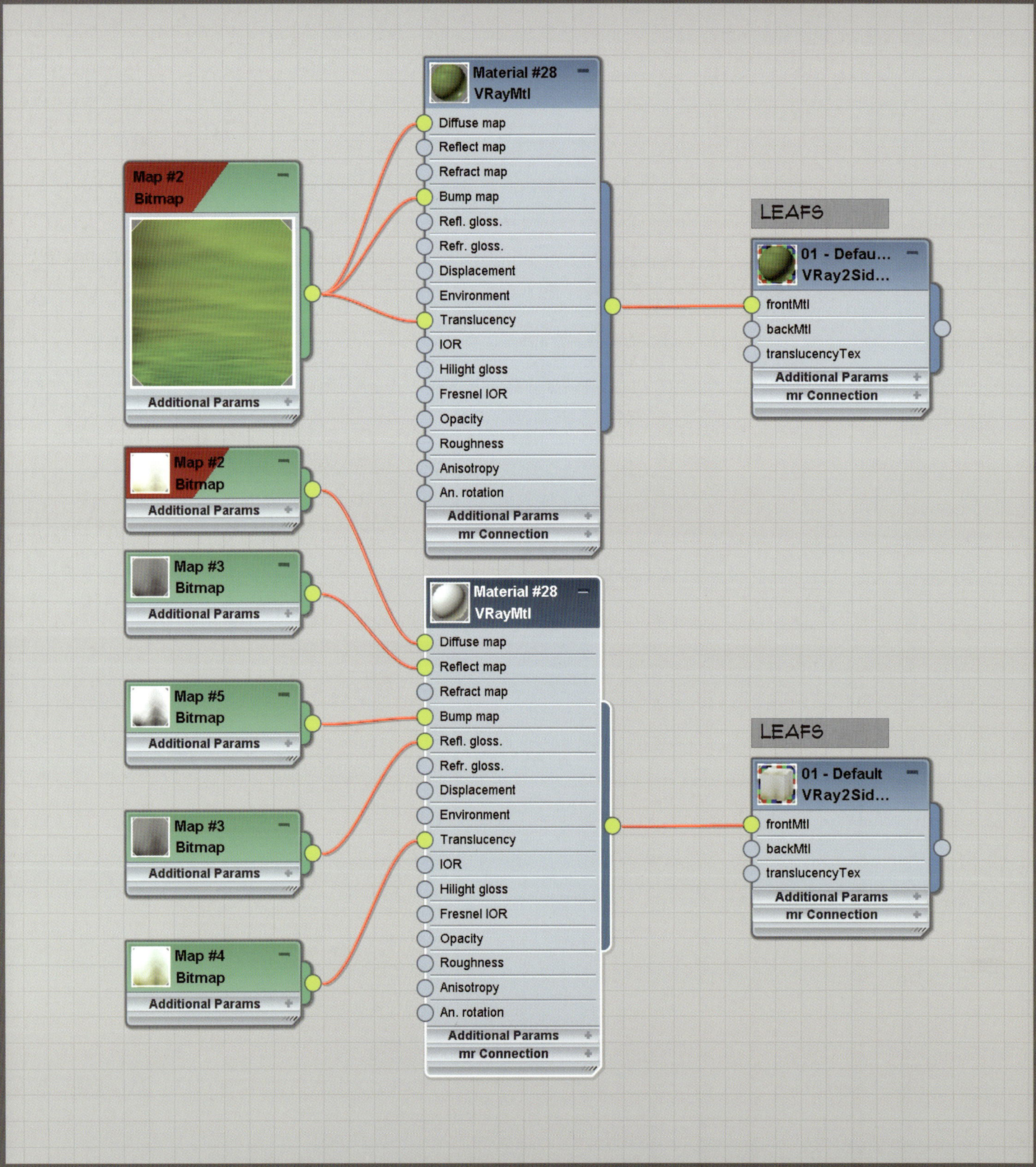

TABLE-WARE VOL I MADE WITH 3DS MAX AND V-RAY

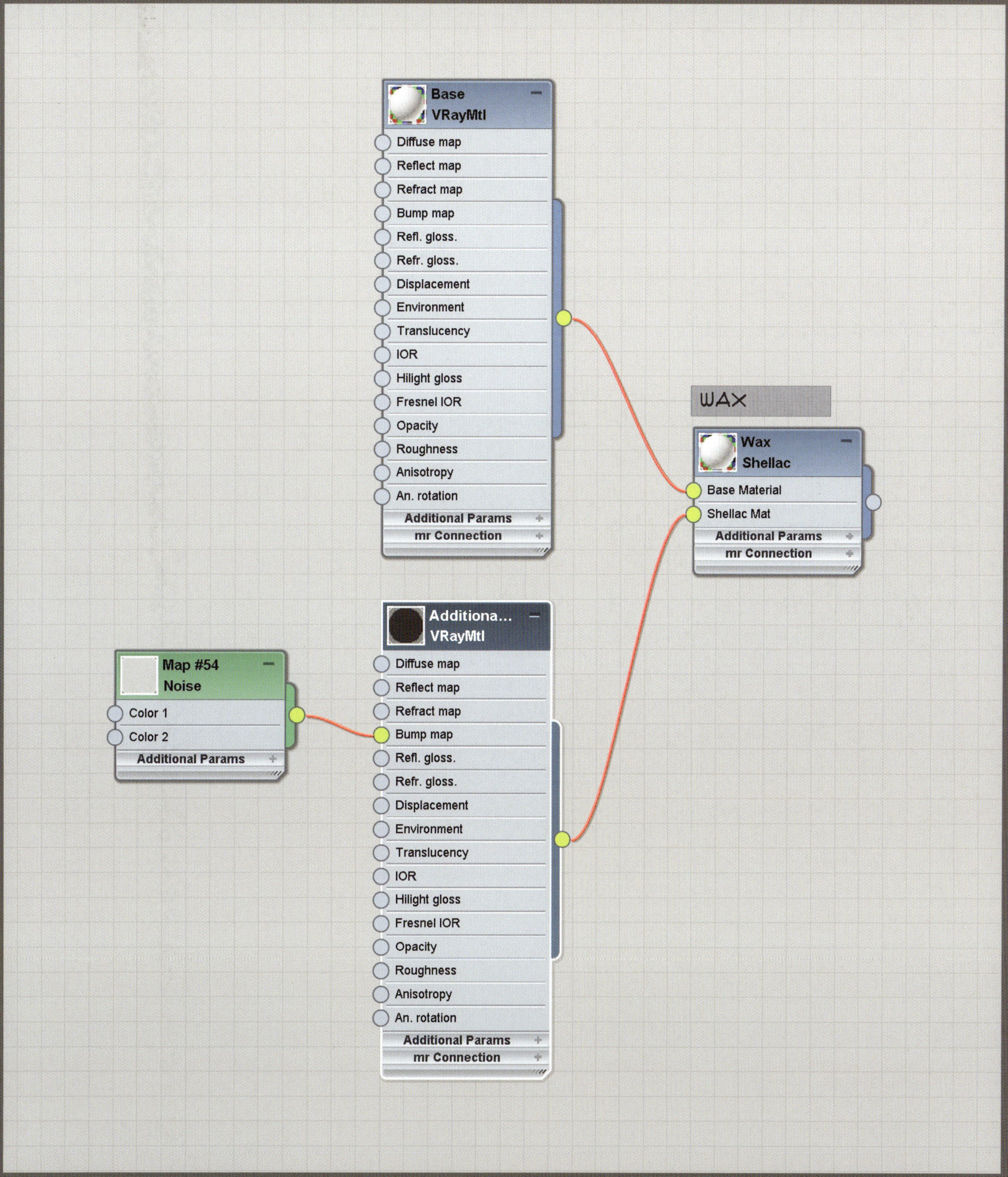

| TABLE-WARE 006 | ⚙ 3DS MAX 2010 + FBX (V-RAY) | 🛍 61,172 KB | ☆ CLASSICAL |

MODIFIERS STACK
- NOT COLLAPSED
- COLLAPSED

TEXTURE
- NOT INCLUDED
- INCLUDED

MAPPING
- UNWRAP
- UVW MAP

TOTAL
- POLYS : 438,577
- VERTS : 392,036

TABLE_WARE

MAX
3DS
FBX - SIMPLE OBJECT WITHOUT MATERIALS (WITH MAPPING AND TEXTURES INCLUDED)
*.MAX - V-RAY 1.5 - OR HIGHER - WITH TEXTURES AND SHADERS
V-RAY - OBJECT PREPARED FOR V-RAY RENDERER (WITH TEXTURES AND SHADERS)

*. MAX 2010 - OR HIGHER

ELEMENT OF MODEL

MAX
3DS
FBX - SIMPLE OBJECT WITHOUT MATERIALS (WITH MAPPING AND TEXTURES INCLUDED)
*.MAX - V-RAY 1.5 - OR HIGHER - WITH TEXTURES AND SHADERS
V-RAY - OBJECT PREPARED FOR V-RAY RENDERER (WITH TEXTURES AND SHADERS)

*. MAX 2010 - OR HIGHER

TABLE-WARE VOL 1 MADE WITH 3DS MAX AND V-RAY

SLATE MATERIAL EDITOR

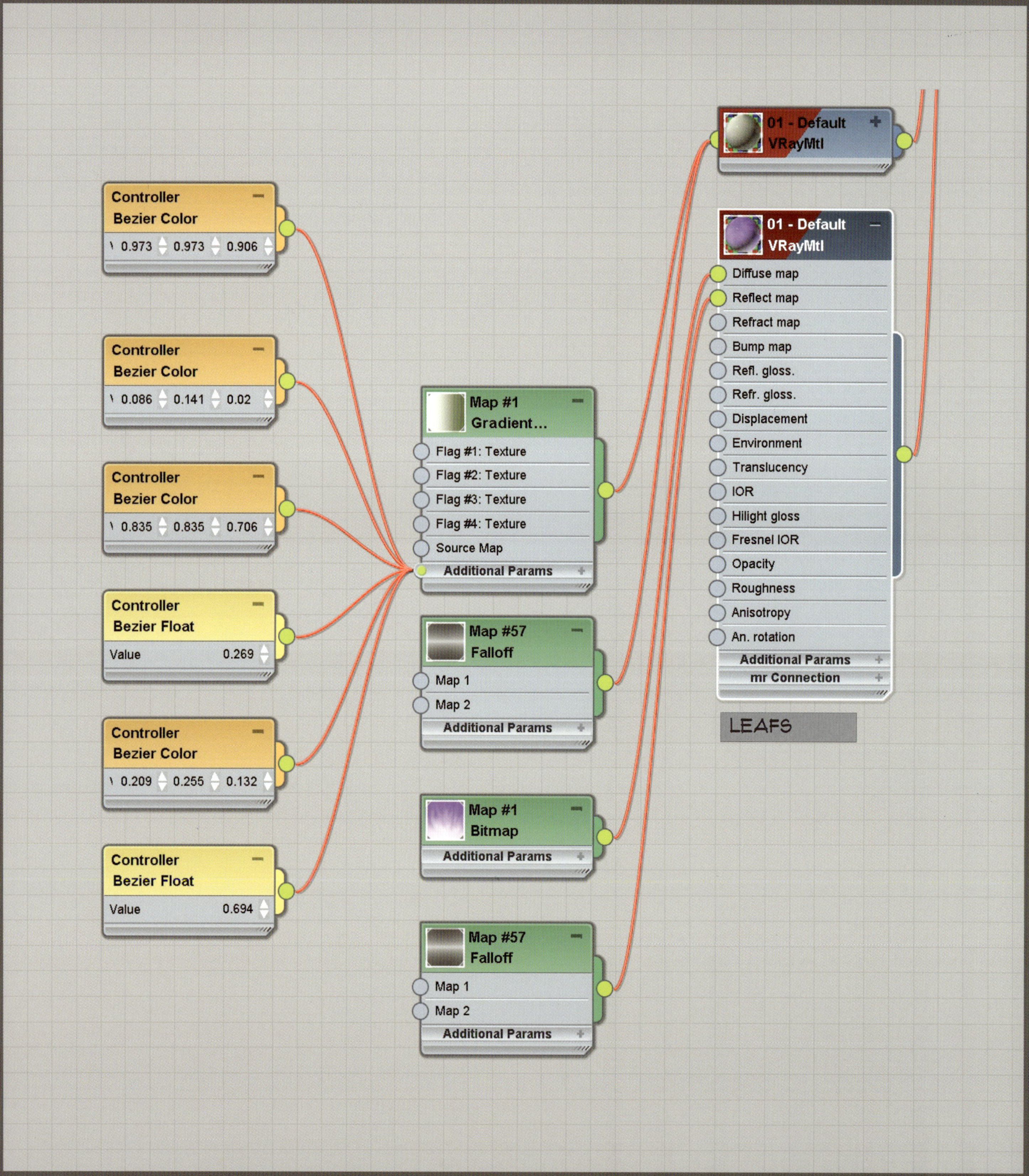

LEAFS

TABLE-WARE VOL I MADE WITH 3DS MAX AND V-RAY

SLATE MATERIAL EDITOR

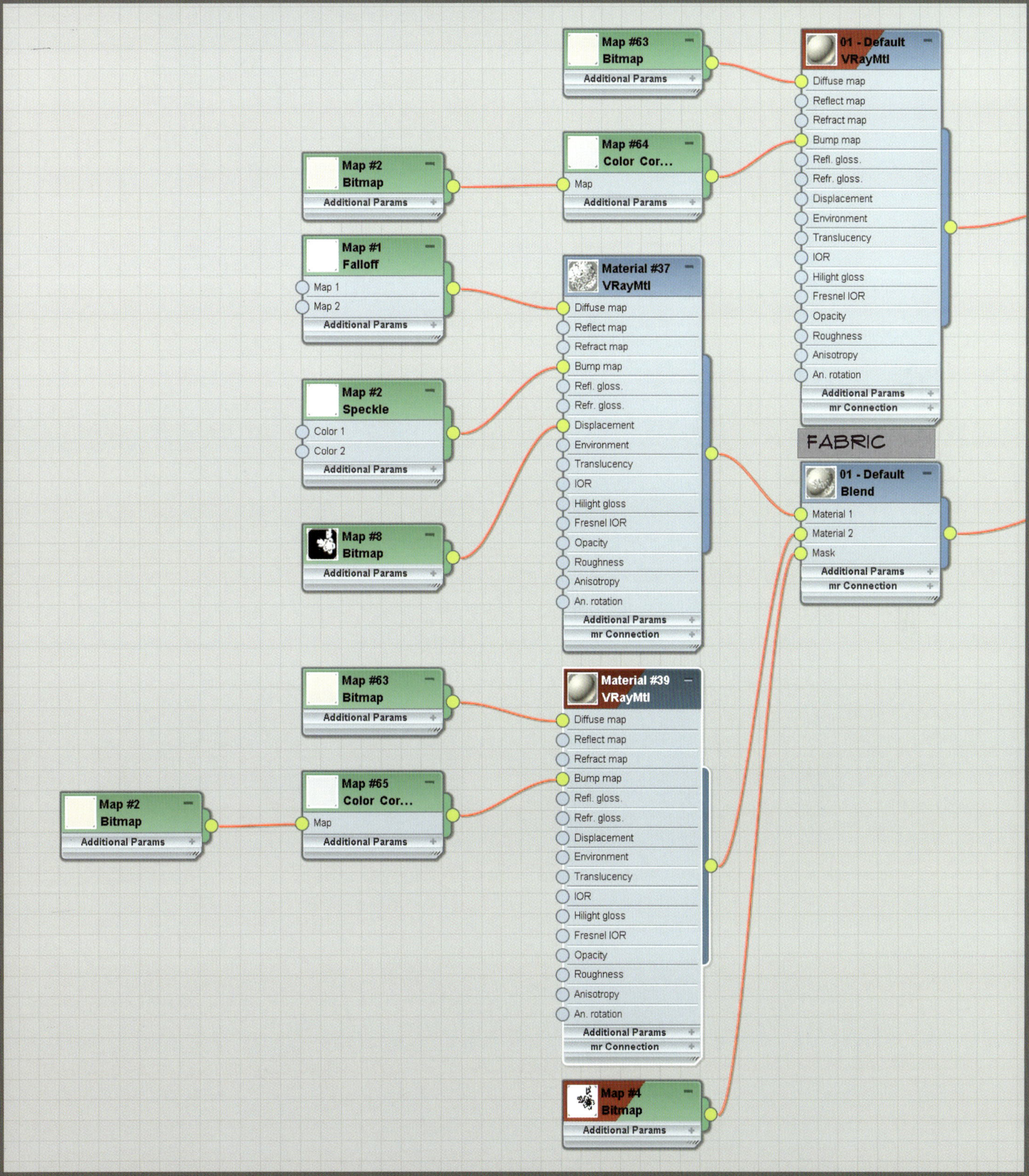

TABLE-WARE VOL I MADE WITH 3DS MAX AND V-RAY

TABLE-WARE 007

3DS MAX 2010 + FBX (V-RAY)   1,944 KB   MODERN

MODIFIERS STACK
NOT COLLAPSED
COLLAPSED

TEXTURE
NOT INCLUDED
INCLUDED

MAPPING
UNWRAP
UVW MAP

TOTAL
POLYS : 28,594
VERTS : 26,711

TABLE_WARE

MAX
3DS
FBX - SIMPLE OBJECT WITHOUT MATERIALS (WITH MAPPING AND TEXTURES INCLUDED)
*.MAX - V-RAY 1.5 - OR HIGHER - WITH TEXTURES AND SHADERS
V-RAY - OBJECT PREPARED FOR V-RAY RENDERER (WITH TEXTURES AND SHADERS)

*. MAX 2010 - OR HIGHER

ELEMENT OF MODEL

PARTIAL GROUP

1. TABLE_WARE-A   2. TABLE_WARE-B   3. TABLE_WARE-C   4. TABLE_WARE-D

TABLE-WARE VOL I MADE WITH 3DS MAX AND V-RAY

SLATE MATERIAL EDITOR

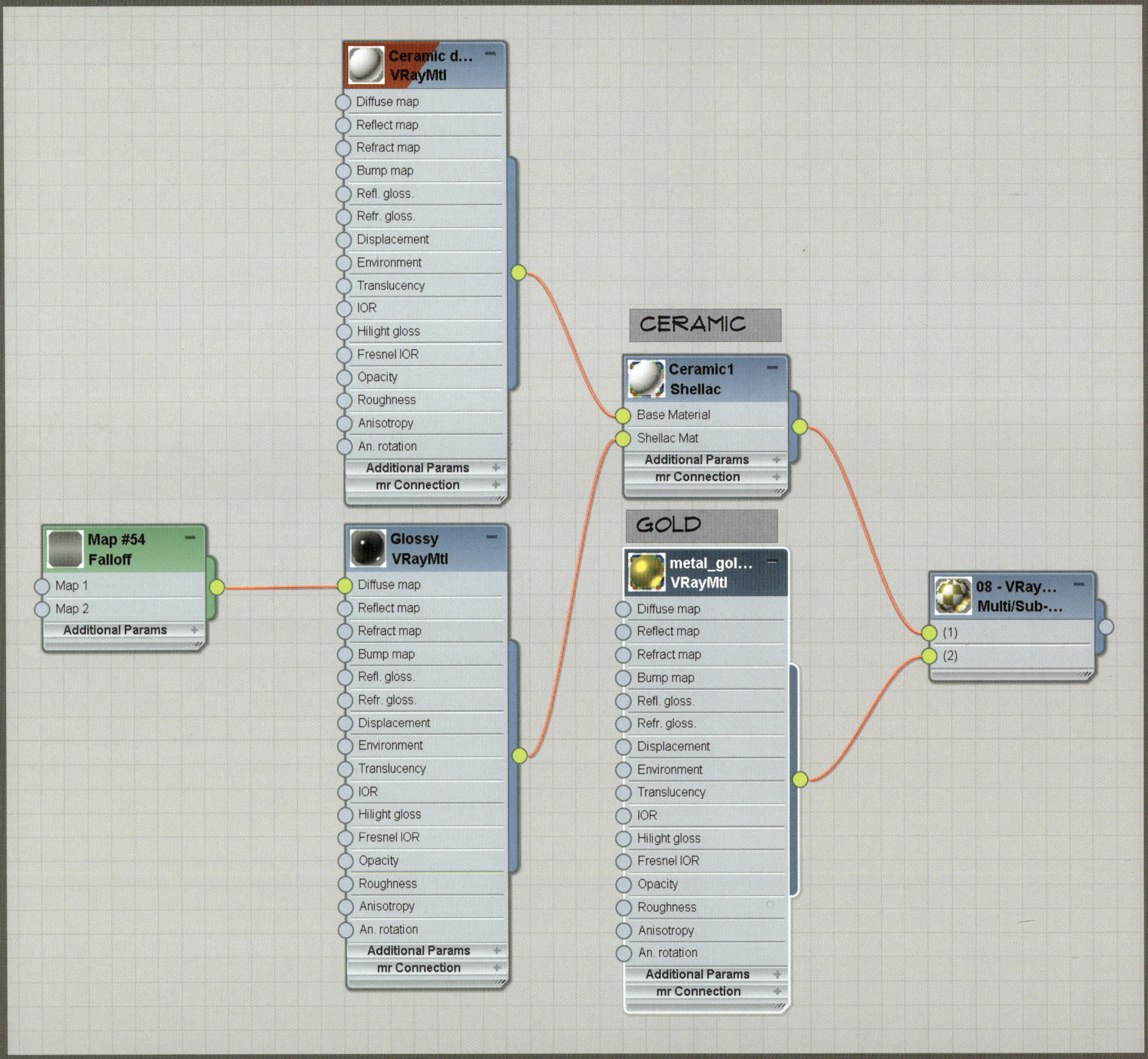

TABLE-WARE VOL 1 MADE WITH 3DS MAX AND V-RAY

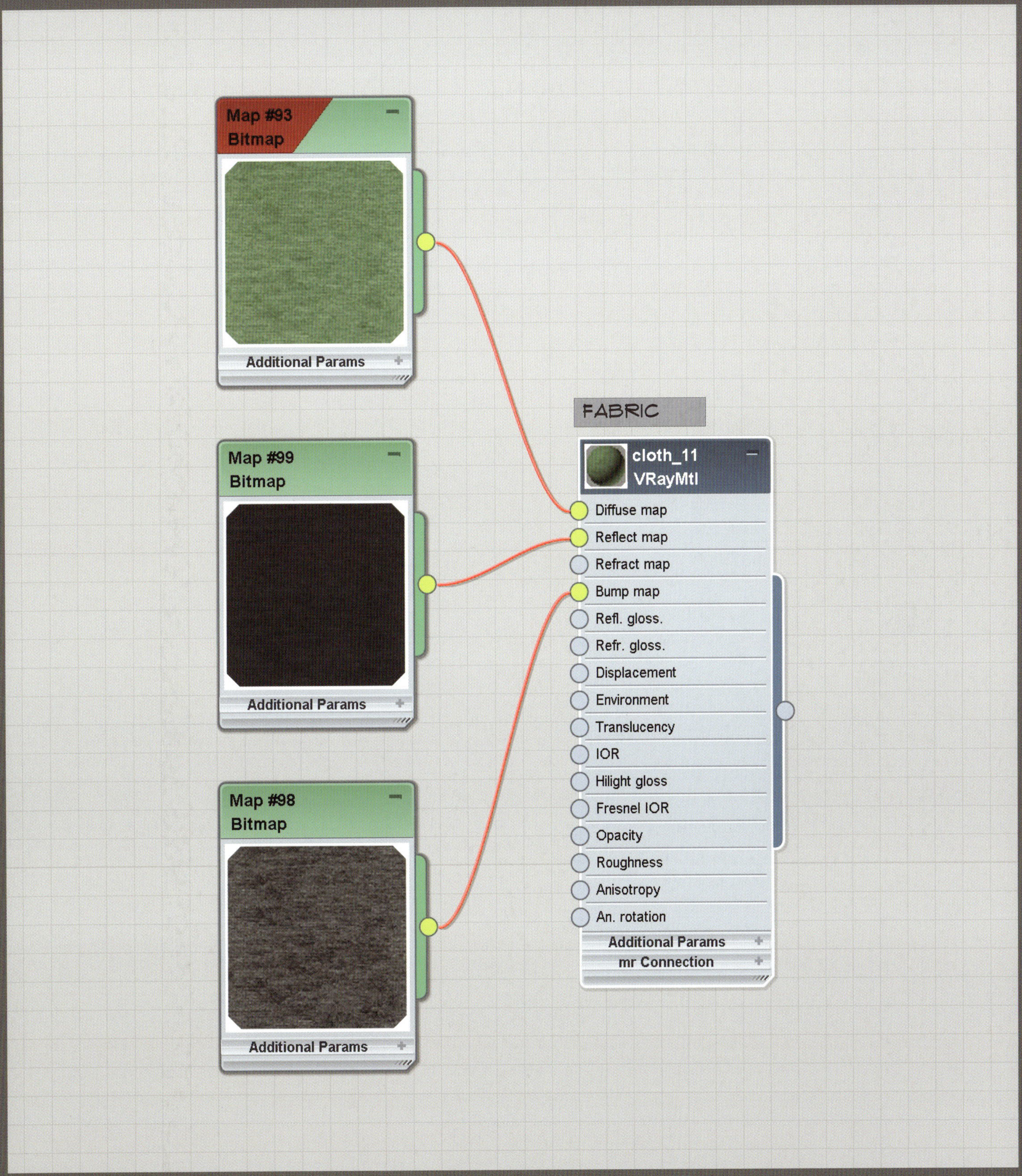

TABLE-WARE 008

3DS MAX 2010 + FBX (V-RAY)   36,080 KB   CLASSICAL

MODIFIERS STACK
- NOT COLLAPSED
- COLLAPSED

TEXTURE
- NOT INCLUDED
- INCLUDED

MAPPING
- UNWRAP
- UVW MAP

TOTAL
- POLYS : 239,940
- VERTS : 249,925

TABLE_WARE

MAX
3DS
FBX – SIMPLE OBJECT WITHOUT MATERIALS (WITH MAPPING AND TEXTURES INCLUDED)
*.MAX – V-RAY 1.5 – OR HIGHER – WITH TEXTURES AND SHADERS
V-RAY – OBJECT PREPARED FOR V-RAY RENDERER (WITH TEXTURES AND SHADERS)

*. MAX 2010 – OR HIGHER

MADE WITH: THE NEW DIGITAL AGE

ELEMENT OF MODEL

PARTIAL GROUP

1. TABLE_WARE-A
2. TABLE_WARE-B
3. TABLE_WARE-C
4. TABLE_WARE-D

TABLE-WARE VOL 1 MADE WITH 3DS MAX AND V-RAY

SLATE MATERIAL EDITOR

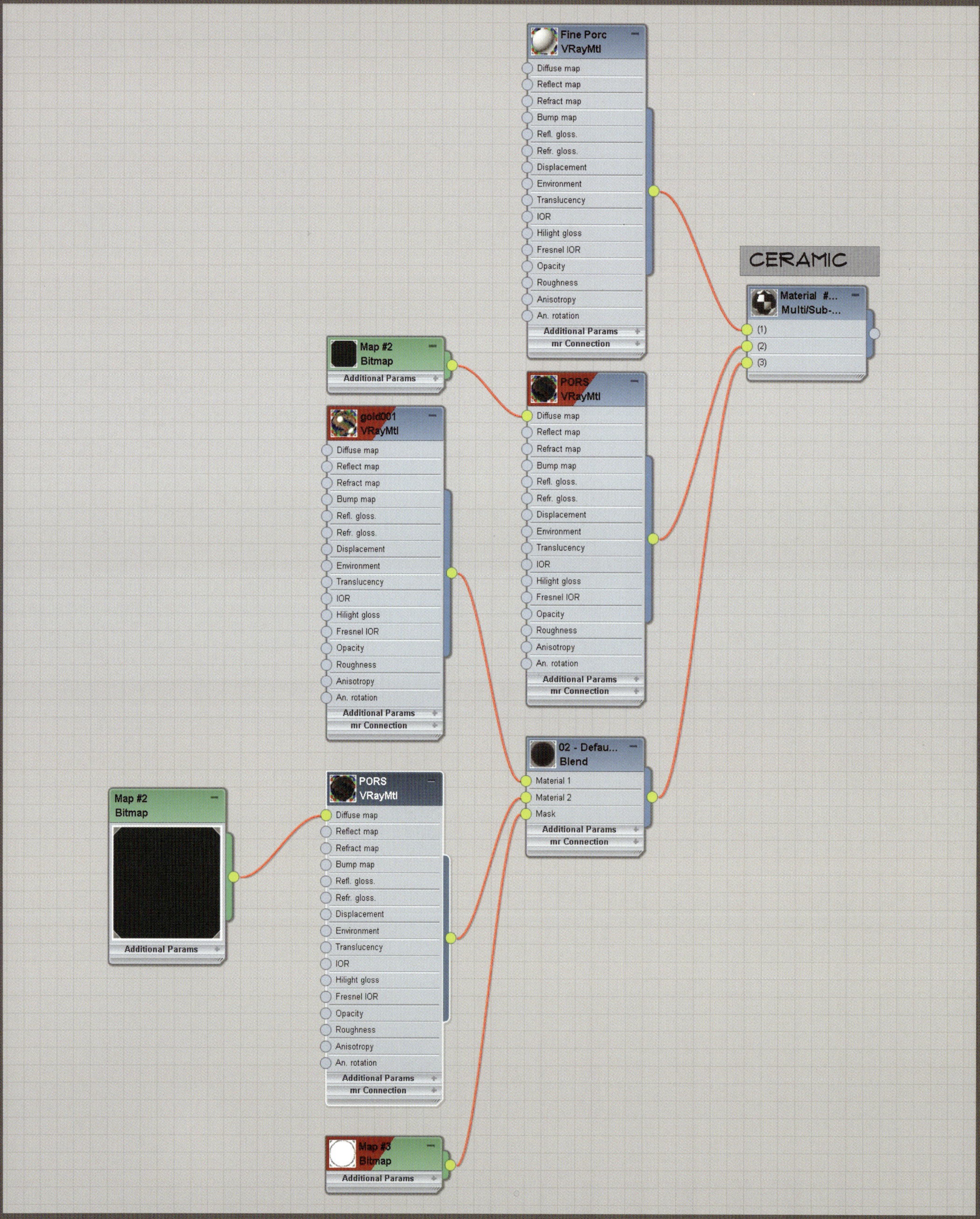

TABLE-WARE VOL I MADE WITH 3DS MAX AND V-RAY

SLATE MATERIAL EDITOR

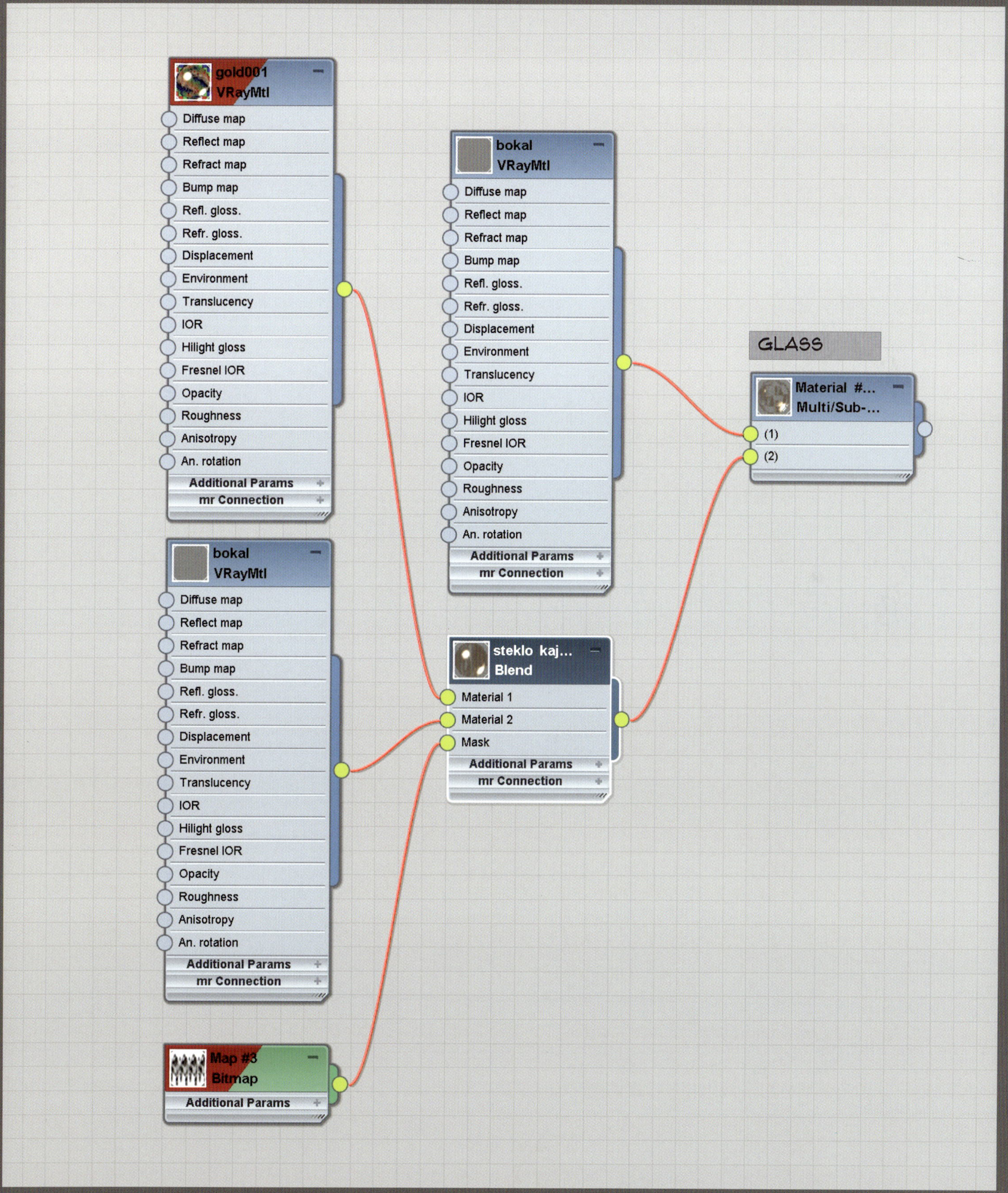

TABLE-WARE VOL I MADE WITH 3DS MAX AND V-RAY

TABLE-WARE 009

3DS MAX 2010 + FBX (V-RAY)   2,404 KB   CLASSICAL

MODIFIERS STACK
NOT COLLAPSED
COLLAPSED

TEXTURE
NOT INCLUDED
INCLUDED

MAPPING
UNWRAP
UVW MAP

TOTAL
POLYS : 82,661
VERTS : 47,612

TABLE_WARE

MAX
3DS
FBX – SIMPLE OBJECT WITHOUT MATERIALS (WITH MAPPING AND TEXTURES INCLUDED)
*.MAX – V-RAY 1.5 – OR HIGHER – WITH TEXTURES AND SHADERS
V-RAY – OBJECT PREPARED FOR V-RAY RENDERER (WITH TEXTURES AND SHADERS)

*. MAX 2010 – OR HIGHER

MADE WITH:
THE NEW DIGITAL AGE

TABLE-WARE VOL 1 MADE WITH 3DS MAX AND V-RAY

ELEMENT OF MODEL

PARTIAL GROUP

1. TABLE_WARE-A

2. TABLE_WARE-B

3. TABLE_WARE-C

TABLE-WARE VOL 1 MADE WITH 3DS MAX AND V-RAY

SLATE MATERIAL EDITOR

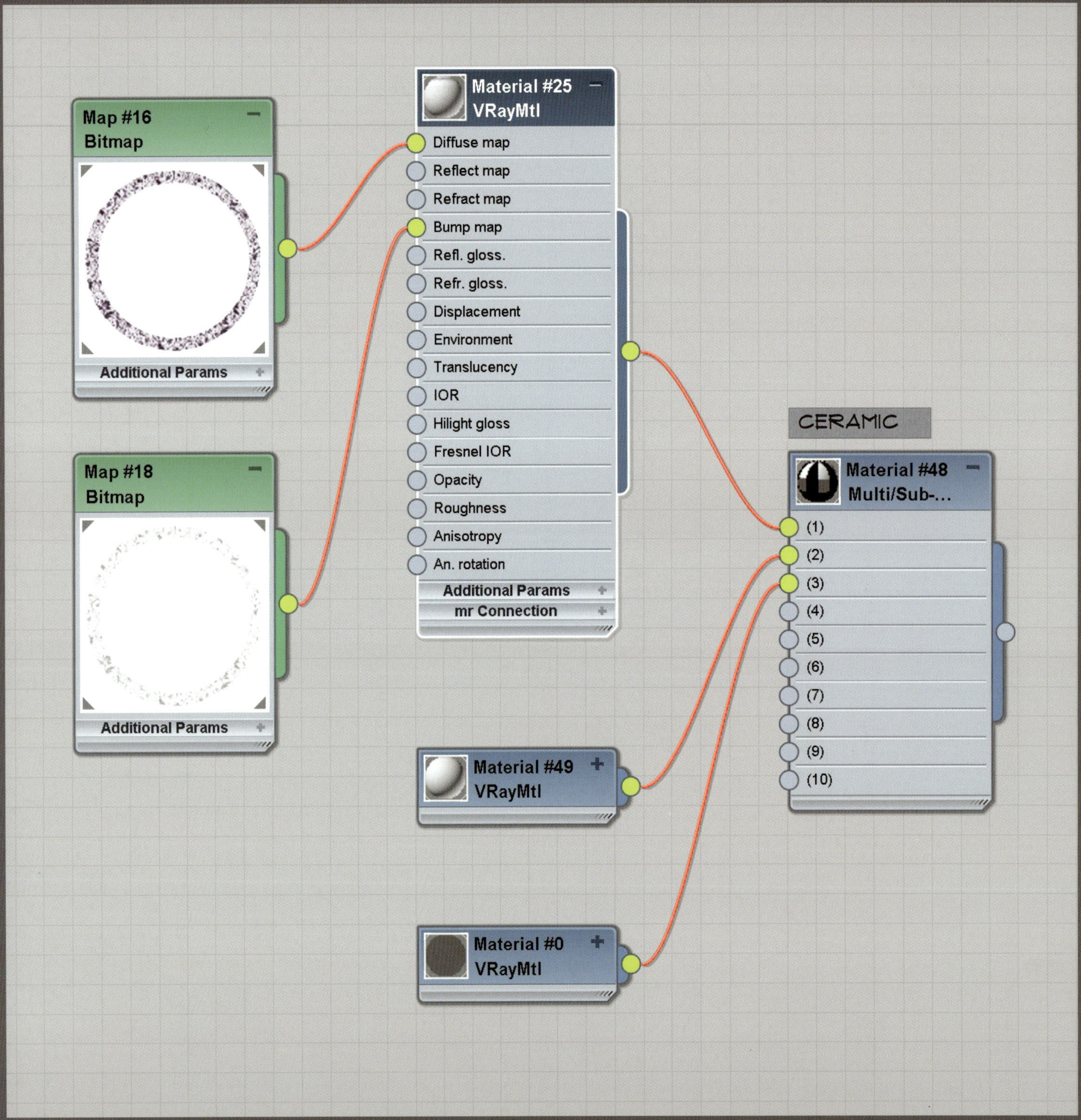

TABLE-WARE VOL I MADE WITH 3DS MAX AND V-RAY

SLATE MATERIAL EDITOR

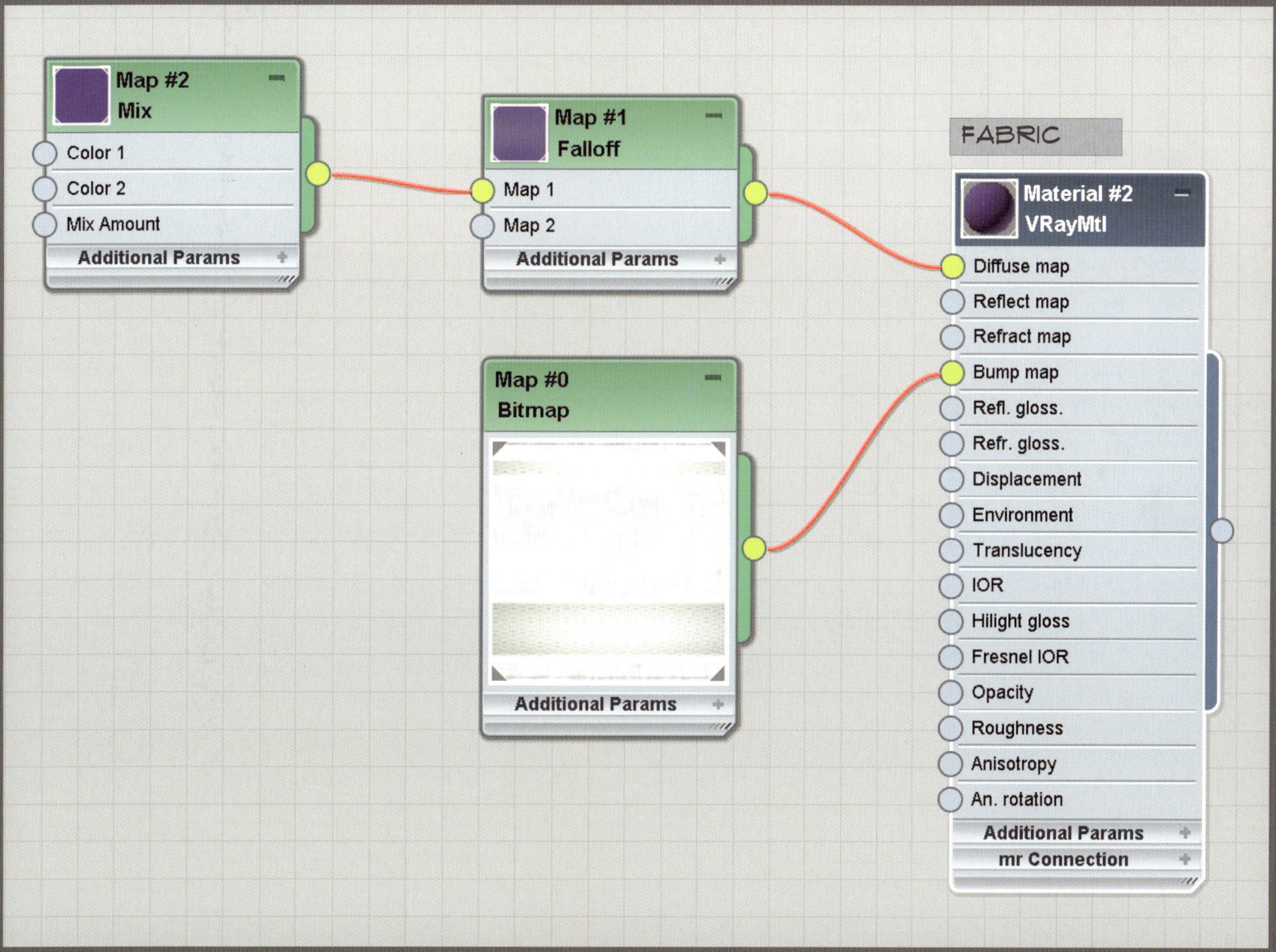

TABLE-WARE VOL 1 MADE WITH 3DS MAX AND V-RAY

| TABLE-WARE 010 | 3DS MAX 2010 + FBX (V-RAY) | 36,440 KB | CLASSICAL |

MODIFIERS STACK
- NOT COLLAPSED
- COLLAPSED

TEXTURE
- NOT INCLUDED
- INCLUDED

MAPPING
- UNWRAP
- UVW MAP

TOTAL
- POLYS : 638,902
- VERTS : 412,603

TABLE_WARE

MAX

3DS

FBX - SIMPLE OBJECT WITHOUT MATERIALS (WITH MAPPING AND TEXTURES INCLUDED)

*.MAX - V-RAY 1.5 - OR HIGHER - WITH TEXTURES AND SHADERS

V-RAY - OBJECT PREPARED FOR V-RAY RENDERER (WITH TEXTURES AND SHADERS)

*. MAX 2010 - OR HIGHER

ELEMENT OF MODEL

PARTIAL GROUP

1. TABLE_WARE-A
2. TABLE_WARE-B
3. TABLE_WARE-C
4. TABLE_WARE-D

TABLE-WARE VOL 1 MADE WITH 3DS MAX AND V-RAY

SLATE MATERIAL EDITOR

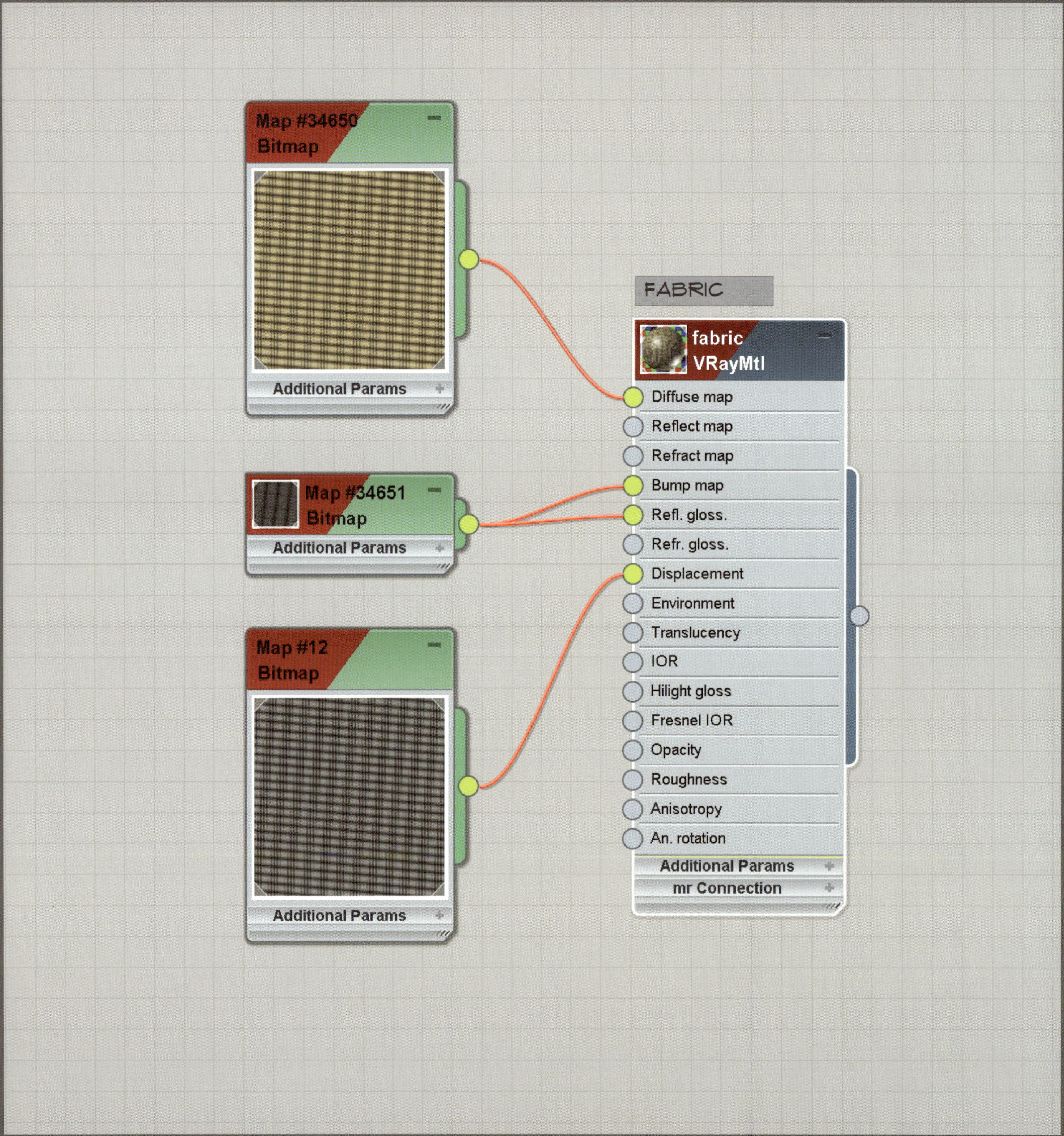

TABLE-WARE VOL I MADE WITH 3DS MAX AND V-RAY

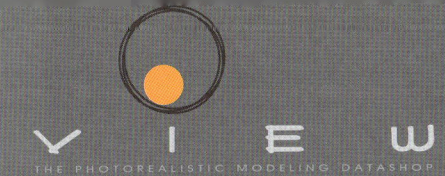

SLATE MATERIAL EDITOR

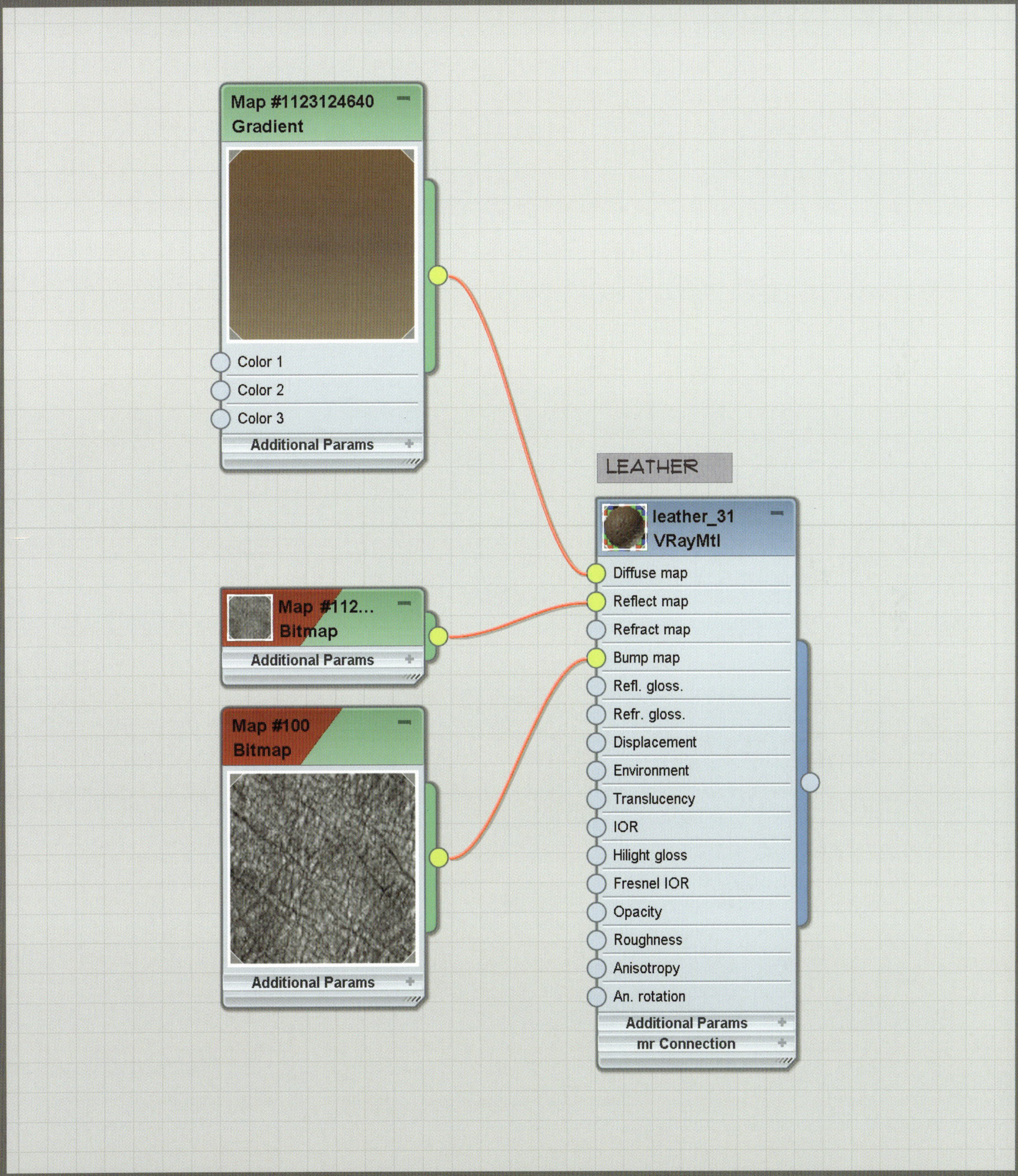

LEATHER

TABLE-WARE VOL I MADE WITH 3DS MAX AND V-RAY

TABLE-WARE 011

3DS MAX 2010 + FBX (V-RAY)   20,384 KB   CLASSICAL

MODIFIERS STACK
- NOT COLLAPSED
- COLLAPSED

TEXTURE
- NOT INCLUDED
- INCLUDED

MAPPING
- UNWRAP
- UVW MAP

TOTAL
- POLYS : 166,785
- VERTS : 163,205

TABLE_WARE

MAX

3DS

FBX - SIMPLE OBJECT WITHOUT MATERIALS (WITH MAPPING AND TEXTURES INCLUDED)

*.MAX - V-RAY 1.5 - OR HIGHER - WITH TEXTURES AND SHADERS

V-RAY - OBJECT PREPARED FOR V-RAY RENDERER (WITH TEXTURES AND SHADERS)

*. MAX 2010 - OR HIGHER

MADE WITH:
THE NEW DIGITAL AGE

ELEMENT OF MODEL

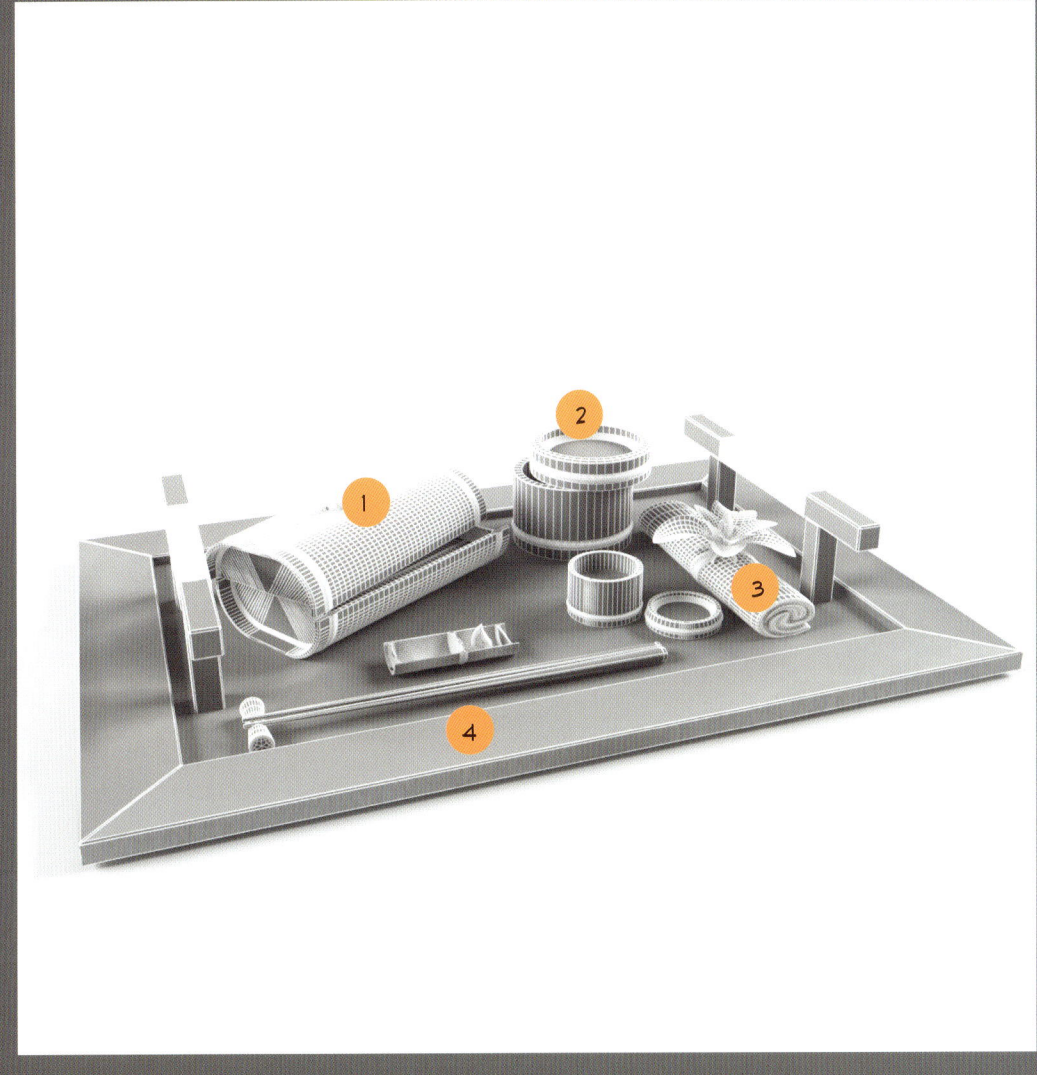

PARTIAL GROUP

1. TABLE_WARE-A

2. TABLE_WARE-B

3. TABLE_WARE-C

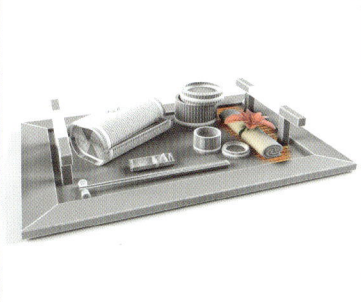

4. TABLE_WARE-D

TABLE-WARE VOL 1 MADE WITH 3DS MAX AND V-RAY

SLATE MATERIAL EDITOR

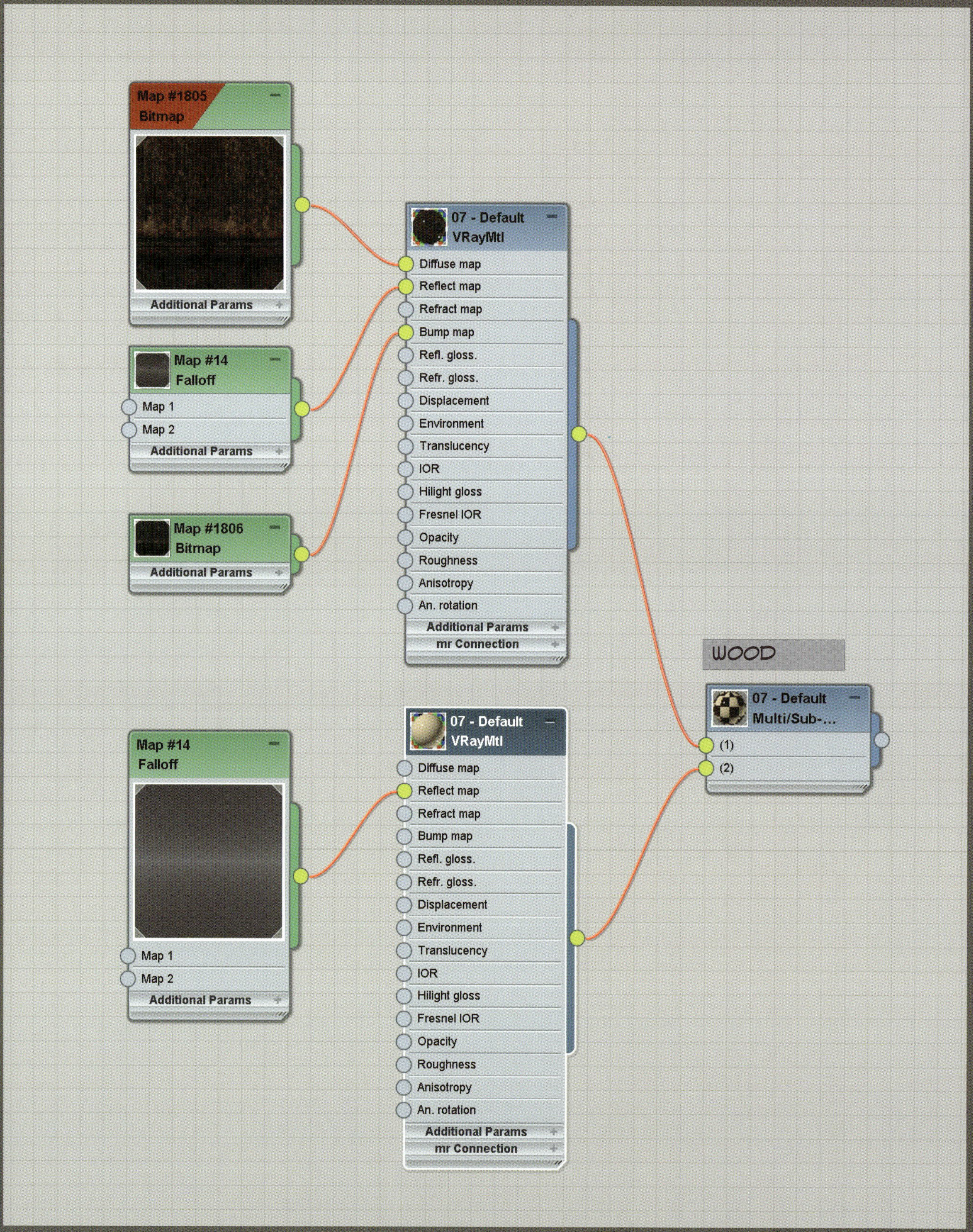

WOOD

SLATE MATERIAL EDITOR

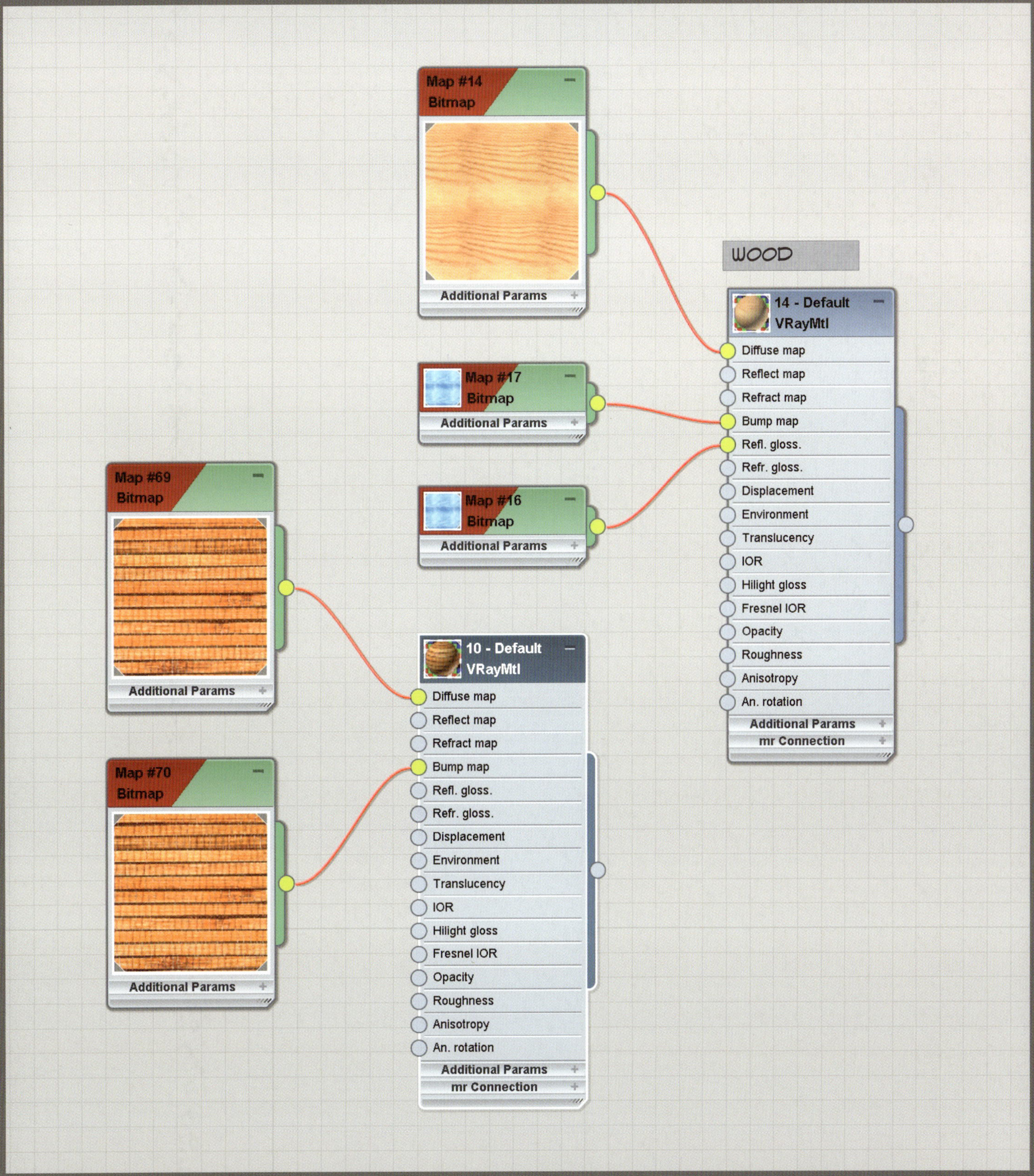

TABLE-WARE 012

3DS MAX 2010 + FBX (V-RAY)   16,488 KB   MODERN

MODIFIERS STACK
- NOT COLLAPSED
- COLLAPSED

TEXTURE
- NOT INCLUDED
- INCLUDED

MAPPING
- UNWRAP
- UVW MAP

TOTAL
- POLYS : 106,432
- VERTS : 107,453

TABLE_WARE

MAX

3DS

FBX – SIMPLE OBJECT WITHOUT MATERIALS (WITH MAPPING AND TEXTURES INCLUDED)

*.MAX – V-RAY 1.5 – OR HIGHER – WITH TEXTURES AND SHADERS

V-RAY – OBJECT PREPARED FOR V-RAY RENDERER (WITH TEXTURES AND SHADERS)

*. MAX 2010 – OR HIGHER

ELEMENT OF MODEL

PARTIAL GROUP

1. TABLE_WARE-A 2. TABLE_WARE-B 3. TABLE_WARE-C 4. TABLE_WARE-D

TABLE-WARE VOL 1 MADE WITH 3DS MAX AND V-RAY

MADE WITH:
THE NEW DIGITAL AGE

SLATE MATERIAL EDITOR

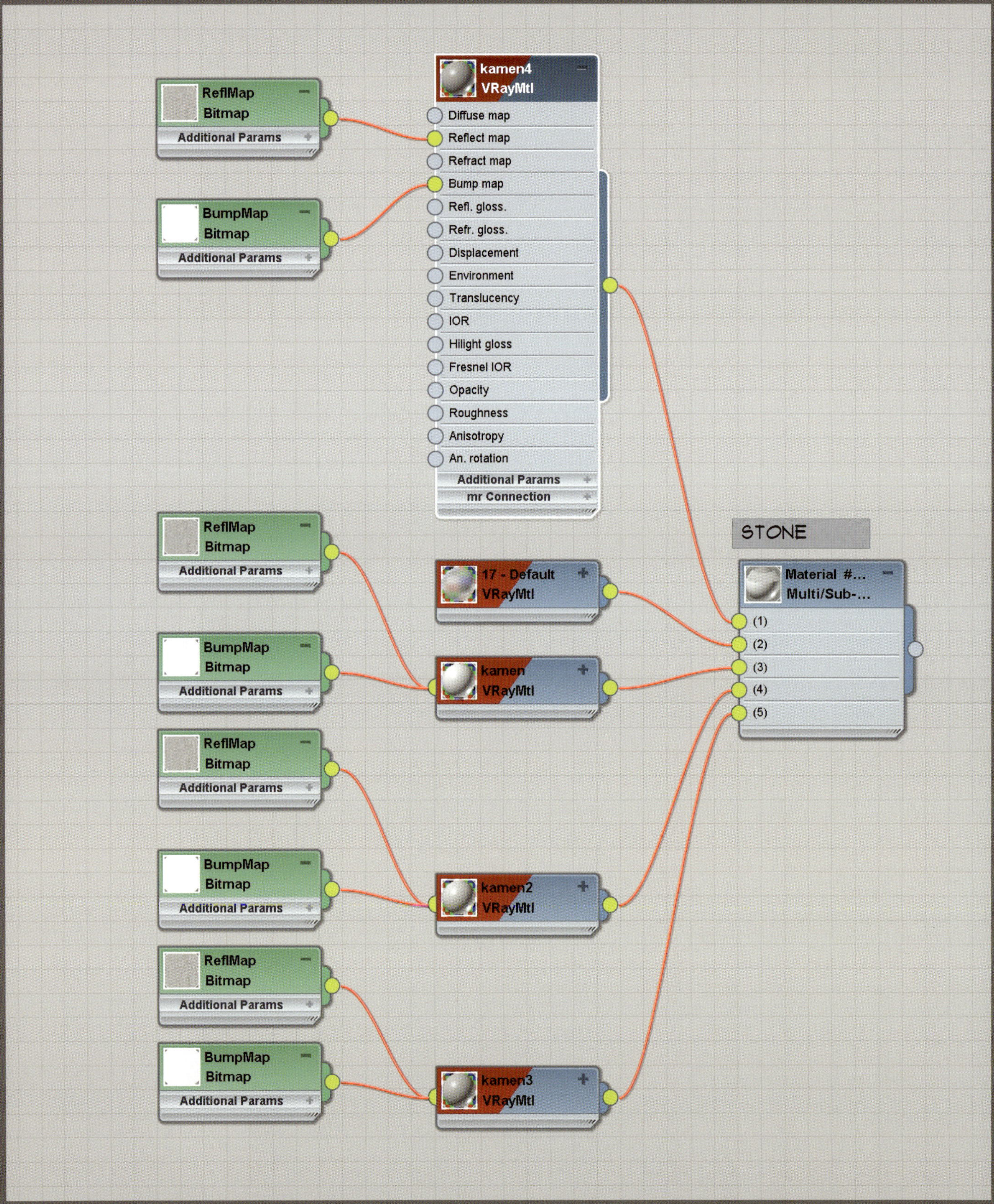

TABLE-WARE VOL I MADE WITH 3DS MAX AND V-RAY

SLATE MATERIAL EDITOR

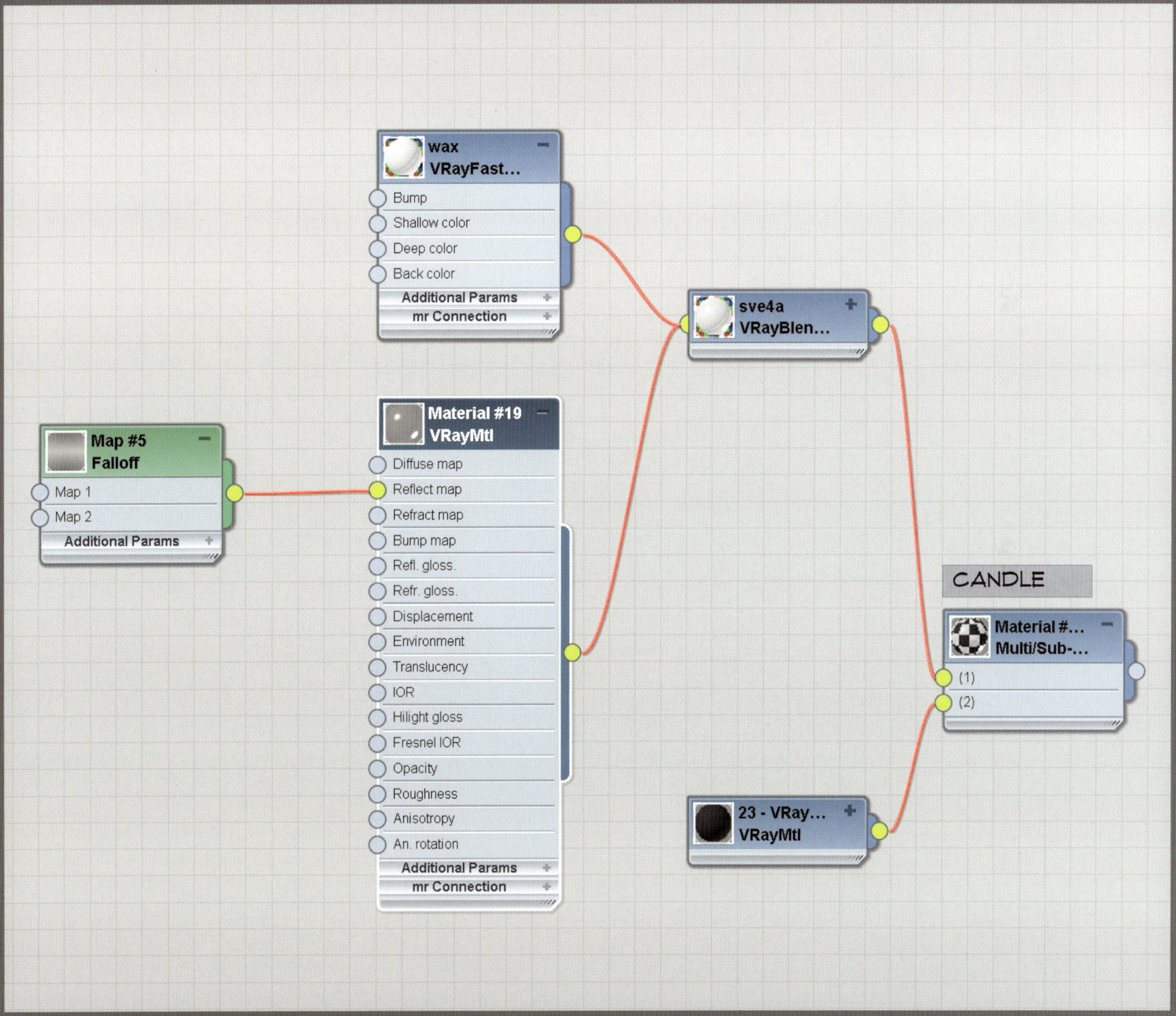

TABLE-WARE VOL 1 MADE WITH 3DS MAX AND V-RAY

TABLE-WARE 013

MODIFIERS STACK
- NOT COLLAPSED
- COLLAPSED

TEXTURE
- NOT INCLUDED
- INCLUDED

MAPPING
- UNWRAP
- UVW MAP

TOTAL
- POLYS : 224,968
- VERTS : 116,018

3DS MAX 2010 + FBX (V-RAY)   4,852 KB   MODERN

TABLE_WARE

MAX

3DS

FBX – SIMPLE OBJECT WITHOUT MATERIALS (WITH MAPPING AND TEXTURES INCLUDED)

*.MAX – V-RAY 1.5 – OR HIGHER – WITH TEXTURES AND SHADERS

V-RAY – OBJECT PREPARED FOR V-RAY RENDERER (WITH TEXTURES AND SHADERS)

*. MAX 2010 – OR HIGHER

ELEMENT OF MODEL

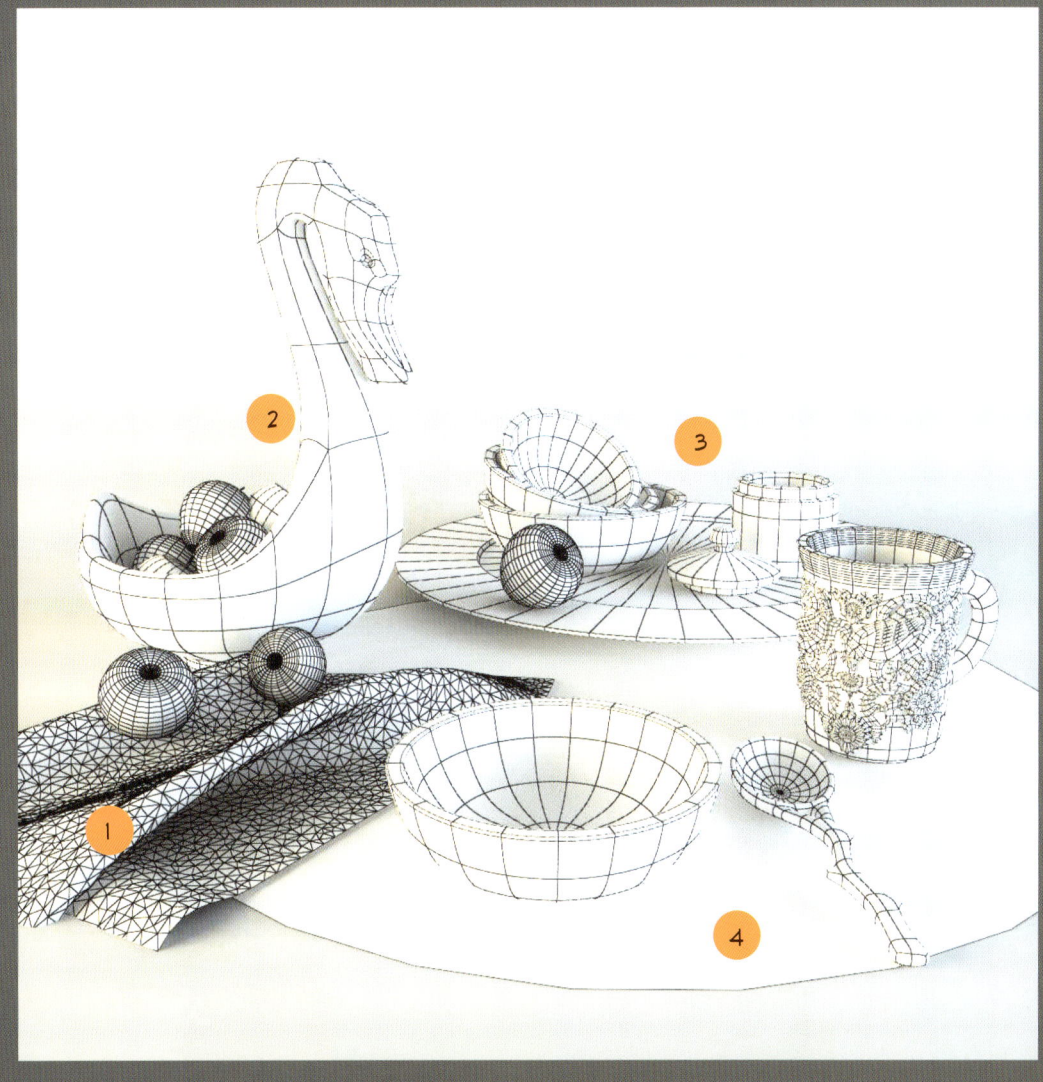

PARTIAL GROUP

1. TABLE_WARE-A 2. TABLE_WARE-B 3. TABLE_WARE-C 4. TABLE_WARE-D

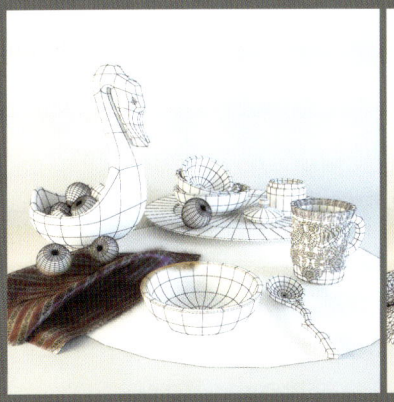

TABLE-WARE VOL 1 MADE WITH 3DS MAX AND V-RAY

SLATE MATERIAL EDITOR

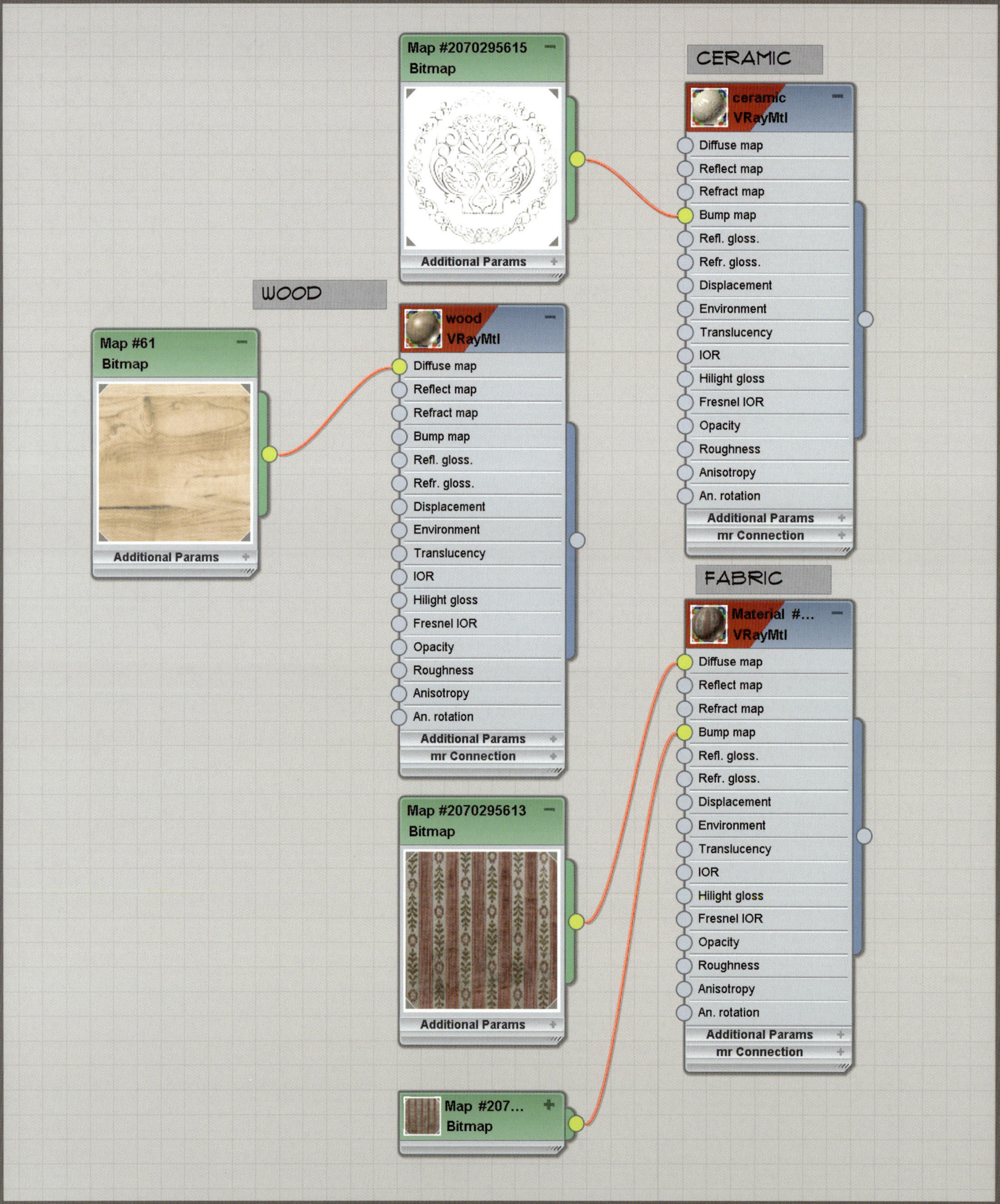

TABLE-WARE VOL 1 MADE WITH 3DS MAX AND V-RAY

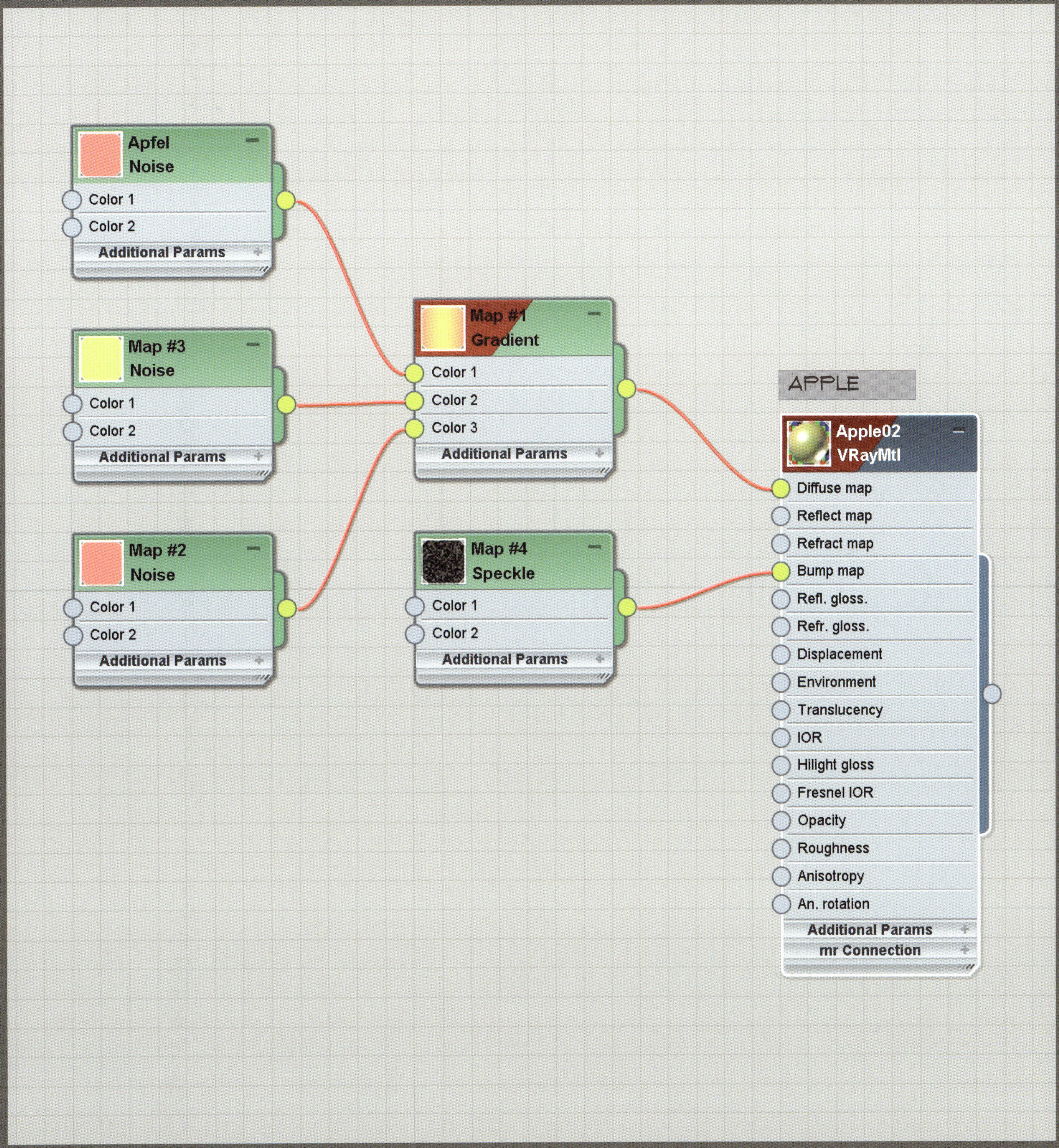

TABLE-WARE 014

3DS MAX 2010 + FBX (V-RAY)   3,580 KB   MODERN

MODIFIERS STACK
NOT COLLAPSED
COLLAPSED

TEXTURE
NOT INCLUDED
INCLUDED

MAPPING
UNWRAP
UVW MAP

TOTAL
POLYS : 502,846
VERTS : 356,975

TABLE_WARE

MAX
3DS
FBX - SIMPLE OBJECT WITHOUT MATERIALS (WITH MAPPING AND TEXTURES INCLUDED)
*.MAX - V-RAY 1.5 - OR HIGHER - WITH TEXTURES AND SHADERS
V-RAY - OBJECT PREPARED FOR V-RAY RENDERER (WITH TEXTURES AND SHADERS)

*. MAX 2010 - OR HIGHER

ELEMENT OF MODEL

PARTIAL GROUP

1. TABLE_WARE-A
2. TABLE_WARE-B
3. TABLE_WARE-C
4. TABLE_WARE-D

TABLE-WARE VOL 1 MADE WITH 3DS MAX AND V-RAY

SLATE MATERIAL EDITOR

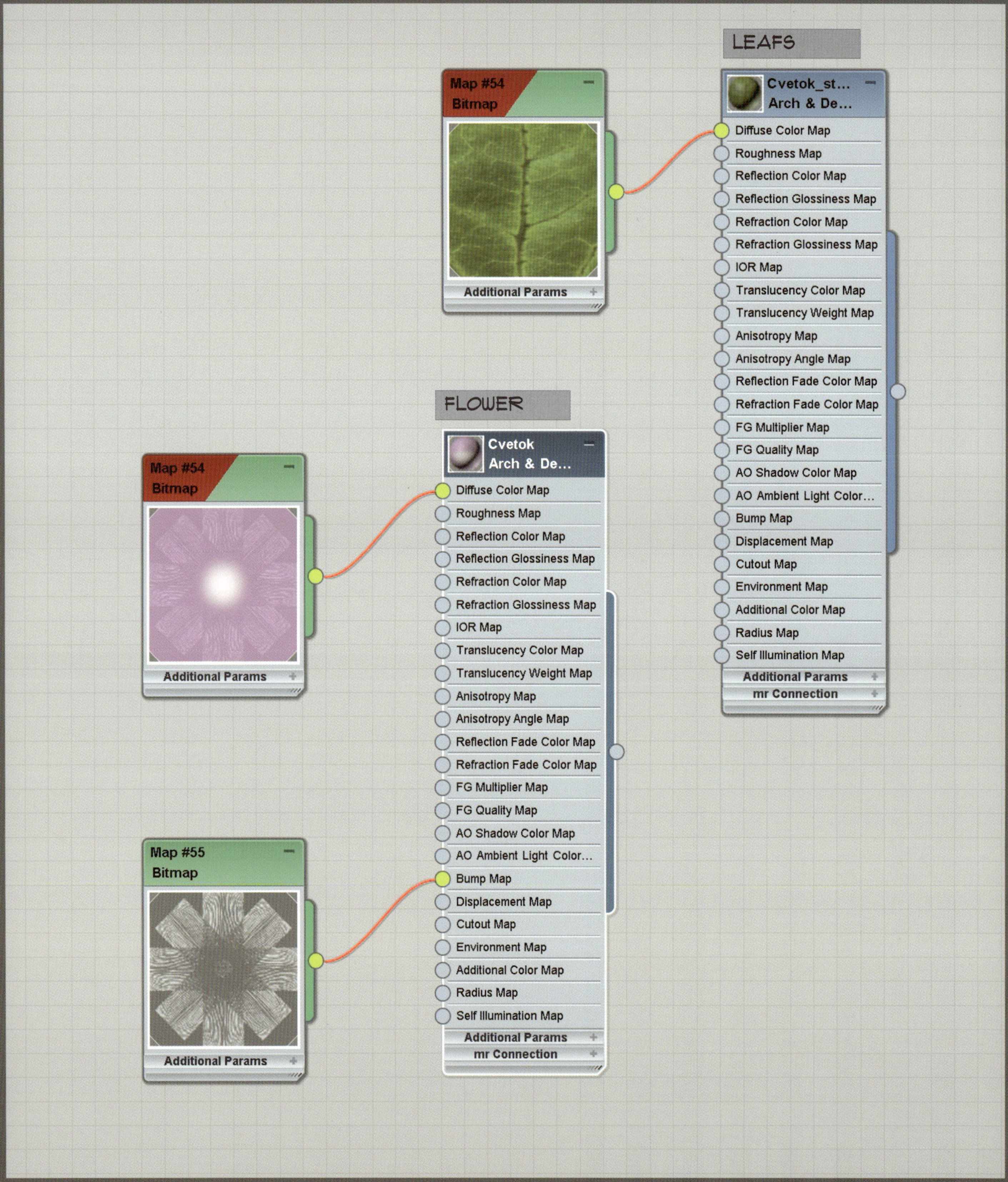

TABLE-WARE VOL I MADE WITH 3DS MAX AND V-RAY

SLATE MATERIAL EDITOR

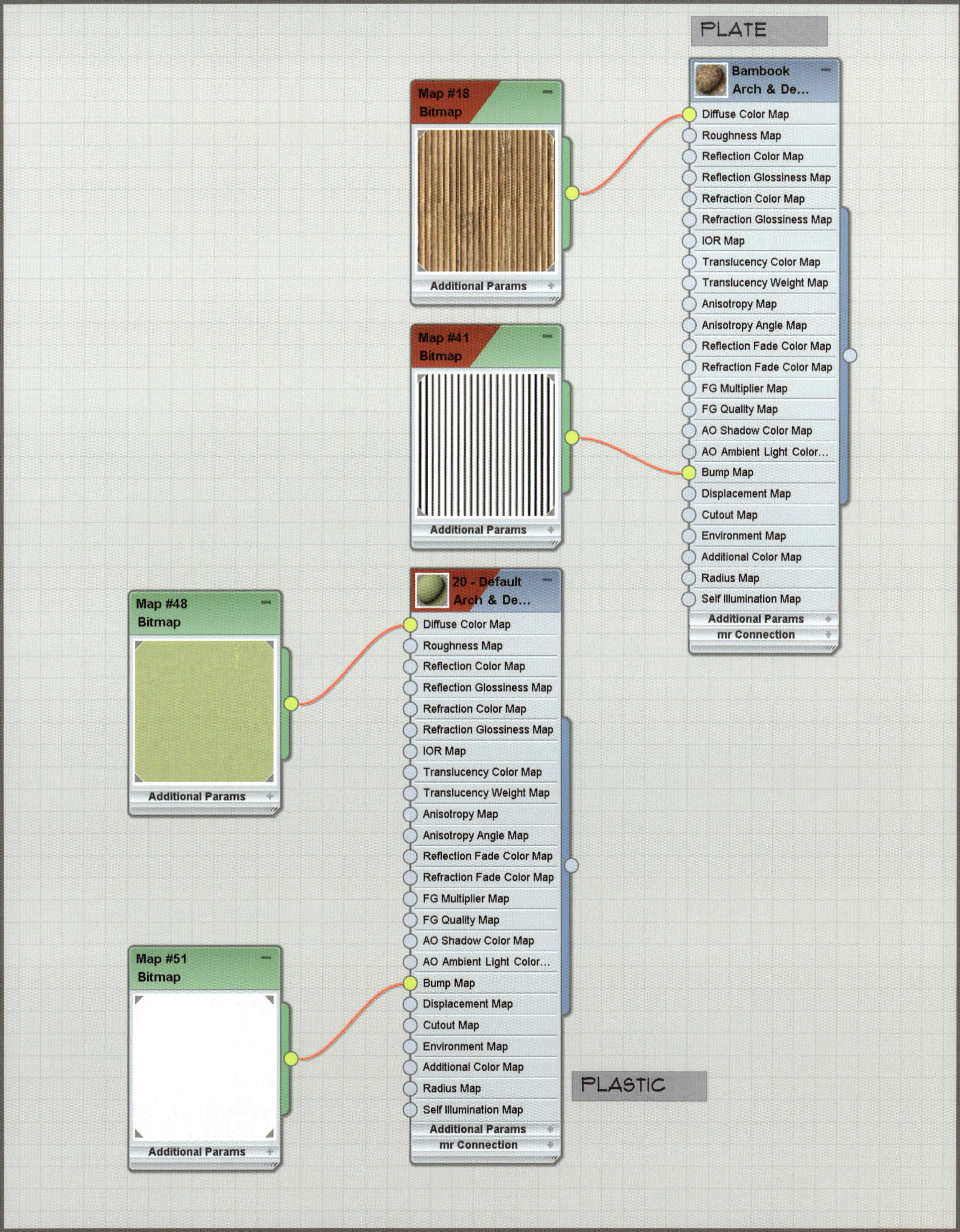

PLATE

PLASTIC

TABLE-WARE VOL 1 MADE WITH 3DS MAX AND V-RAY

TABLE-WARE 015

MODIFIERS STACK
- NOT COLLAPSED
- COLLAPSED

TEXTURE
- NOT INCLUDED
- INCLUDED

MAPPING
- UNWRAP
- UVW MAP

TOTAL
- POLYS : 96,692
- VERTS : 48,601

3DS MAX 2010 + FBX (V-RAY)   5,880 KB   CLASSICAL

TABLE_WARE

MAX
3DS
FBX — SIMPLE OBJECT WITHOUT MATERIALS (WITH MAPPING AND TEXTURES INCLUDED)
*.MAX — V-RAY 1.5 — OR HIGHER — WITH TEXTURES AND SHADERS
V-RAY — OBJECT PREPARED FOR V-RAY RENDERER (WITH TEXTURES AND SHADERS)

*. MAX 2010 — OR HIGHER

MADE WITH: THE NEW DIGITAL AGE

ELEMENT OF MODEL

PARTIAL GROUP

1. TABLE_WARE-A 2. TABLE_WARE-B 3. TABLE_WARE-C 4. TABLE_WARE-D

TABLE-WARE VOL 1 MADE WITH 3DS MAX AND V-RAY

MADE WITH:
THE NEW DIGITAL AGE

SLATE MATERIAL EDITOR

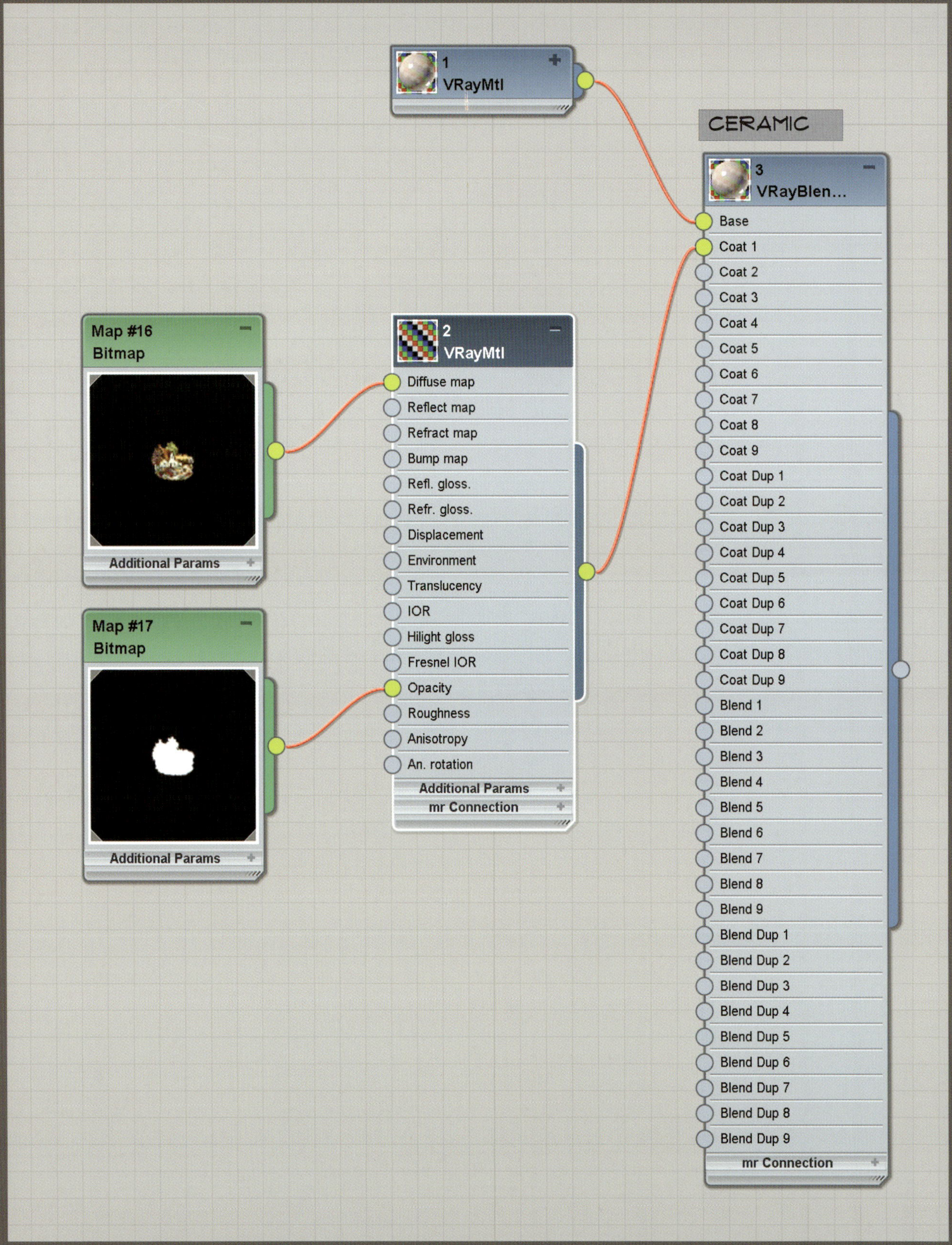

CERAMIC

TABLE-WARE VOL 1 MADE WITH 3DS MAX AND V-RAY

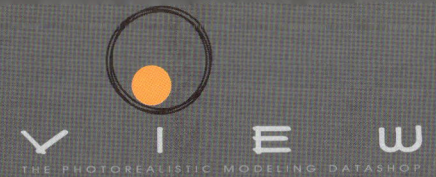

SLATE MATERIAL EDITOR

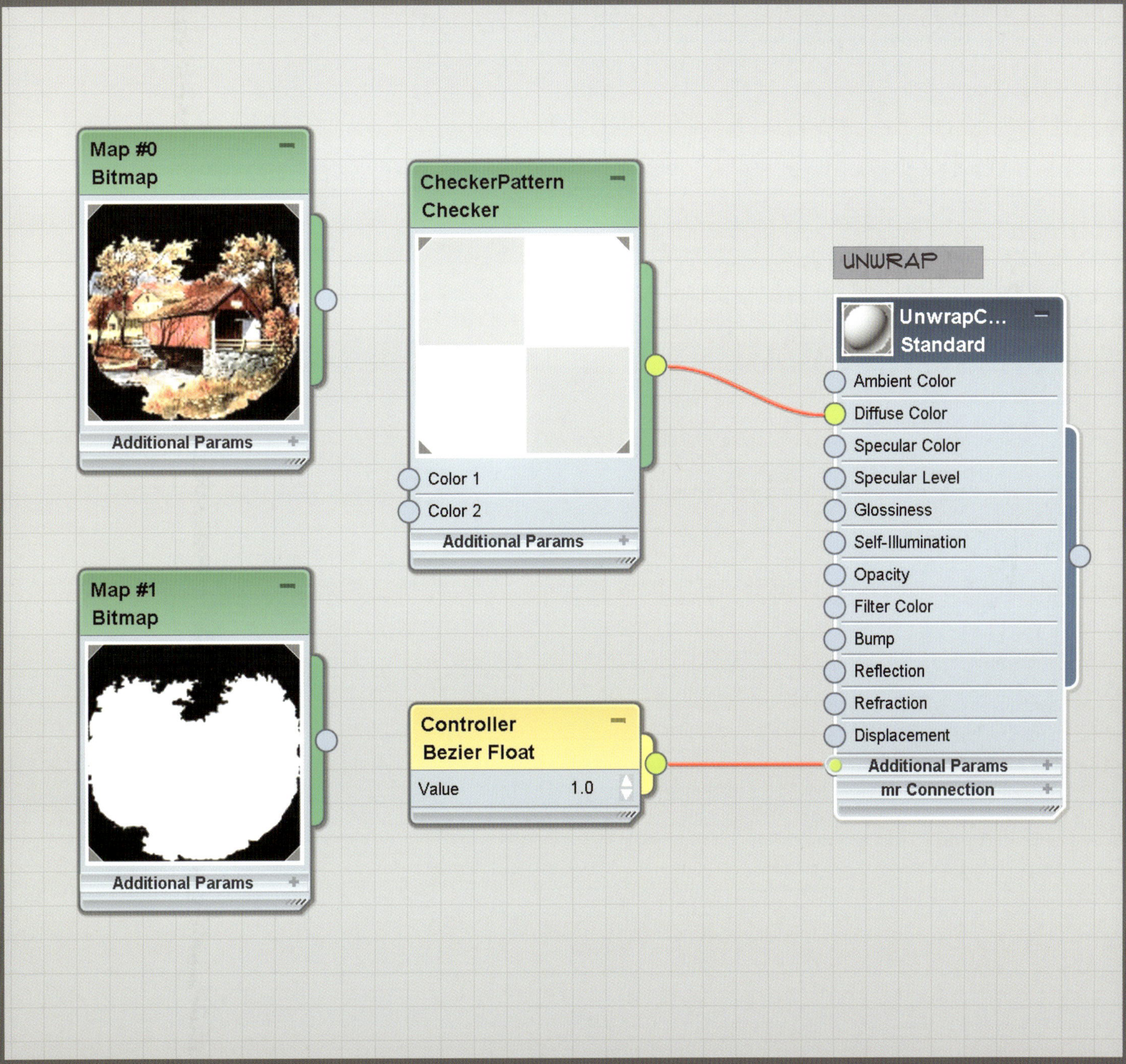

TABLE-WARE VOL I MADE WITH 3DS MAX AND V-RAY

TABLE-WARE 016

3DS MAX 2010 + FBX (V-RAY)   14,508 KB   MODERN

MODIFIERS STACK
- NOT COLLAPSED
- COLLAPSED

TEXTURE
- NOT INCLUDED
- INCLUDED

MAPPING
- UNWRAP
- UVW MAP

TOTAL
- POLYS : 153,312
- VERTS : 122,410

TABLE_WARE

MAX

3DS

FBX - SIMPLE OBJECT WITHOUT MATERIALS (WITH MAPPING AND TEXTURES INCLUDED)

*.MAX - V-RAY 1.5 - OR HIGHER - WITH TEXTURES AND SHADERS

V-RAY - OBJECT PREPARED FOR V-RAY RENDERER (WITH TEXTURES AND SHADERS)

*. MAX 2010 - OR HIGHER

ELEMENT OF MODEL

PARTIAL GROUP

1. TABLE_WARE-A   2. TABLE_WARE-B   3. TABLE_WARE-C   4. TABLE_WARE-D

TABLE-WARE VOL 1 MADE WITH 3DS MAX AND V-RAY

SLATE MATERIAL EDITOR

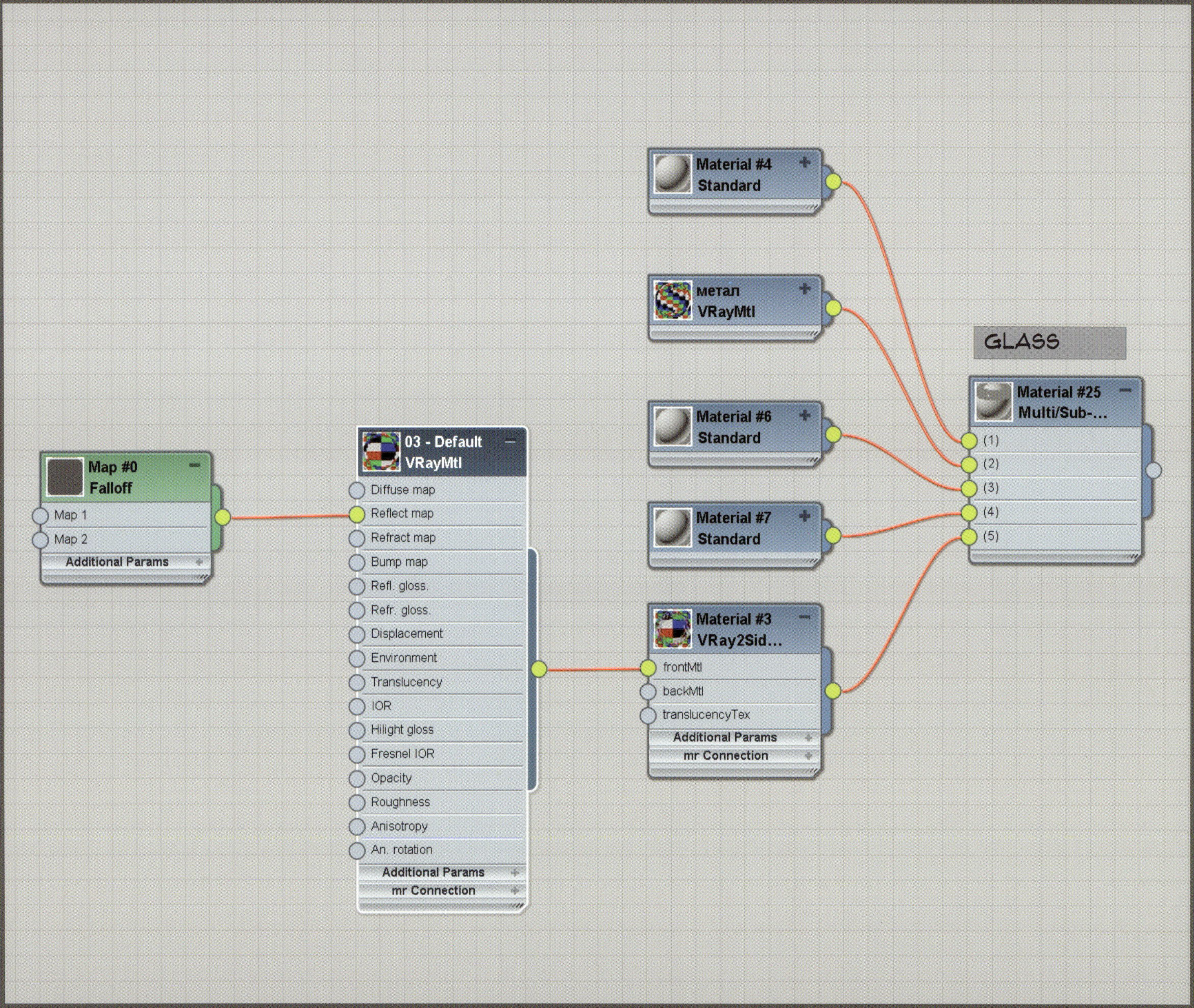

TABLE-WARE VOL I MADE WITH 3DS MAX AND V-RAY

MADE WITH:
THE NEW DIGITAL AGE

SLATE MATERIAL EDITOR

TABLE-WARE VOL I MADE WITH 3DS MAX AND V-RAY

TABLE-WARE 017

3DS MAX 2010 + FBX (V-RAY)   13,600 KB   CLASSICAL

MODIFIERS STACK
- NOT COLLAPSED
- COLLAPSED

TEXTURE
- NOT INCLUDED
- INCLUDED

MAPPING
- UNWRAP
- UVW MAP

TOTAL
POLYS : 106,426
VERTS : 99,966

TABLE_WARE

MAX

3DS

FBX – SIMPLE OBJECT WITHOUT MATERIALS (WITH MAPPING AND TEXTURES INCLUDED)

*.MAX – V-RAY 1.5 – OR HIGHER – WITH TEXTURES AND SHADERS

V-RAY – OBJECT PREPARED FOR V-RAY RENDERER (WITH TEXTURES AND SHADERS)

*. MAX 2010 – OR HIGHER

ELEMENT OF MODEL

PARTIAL GROUP

1. TABLE_WARE-A 2. TABLE_WARE-B 3. TABLE_WARE-C 4. TABLE_WARE-D

TABLE-WARE VOL 1 MADE WITH 3DS MAX AND V-RAY

SLATE MATERIAL EDITOR

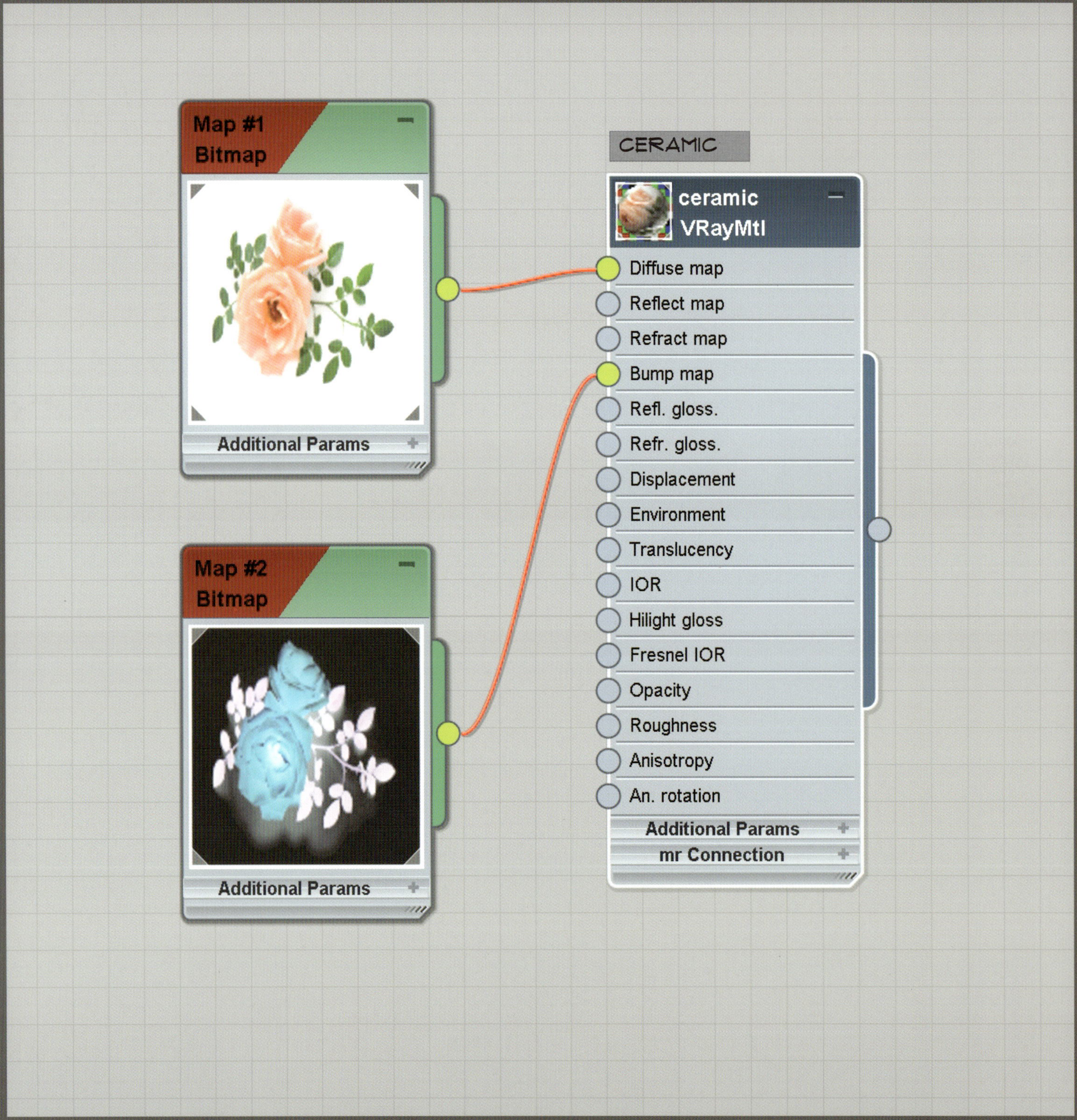

SLATE MATERIAL EDITOR

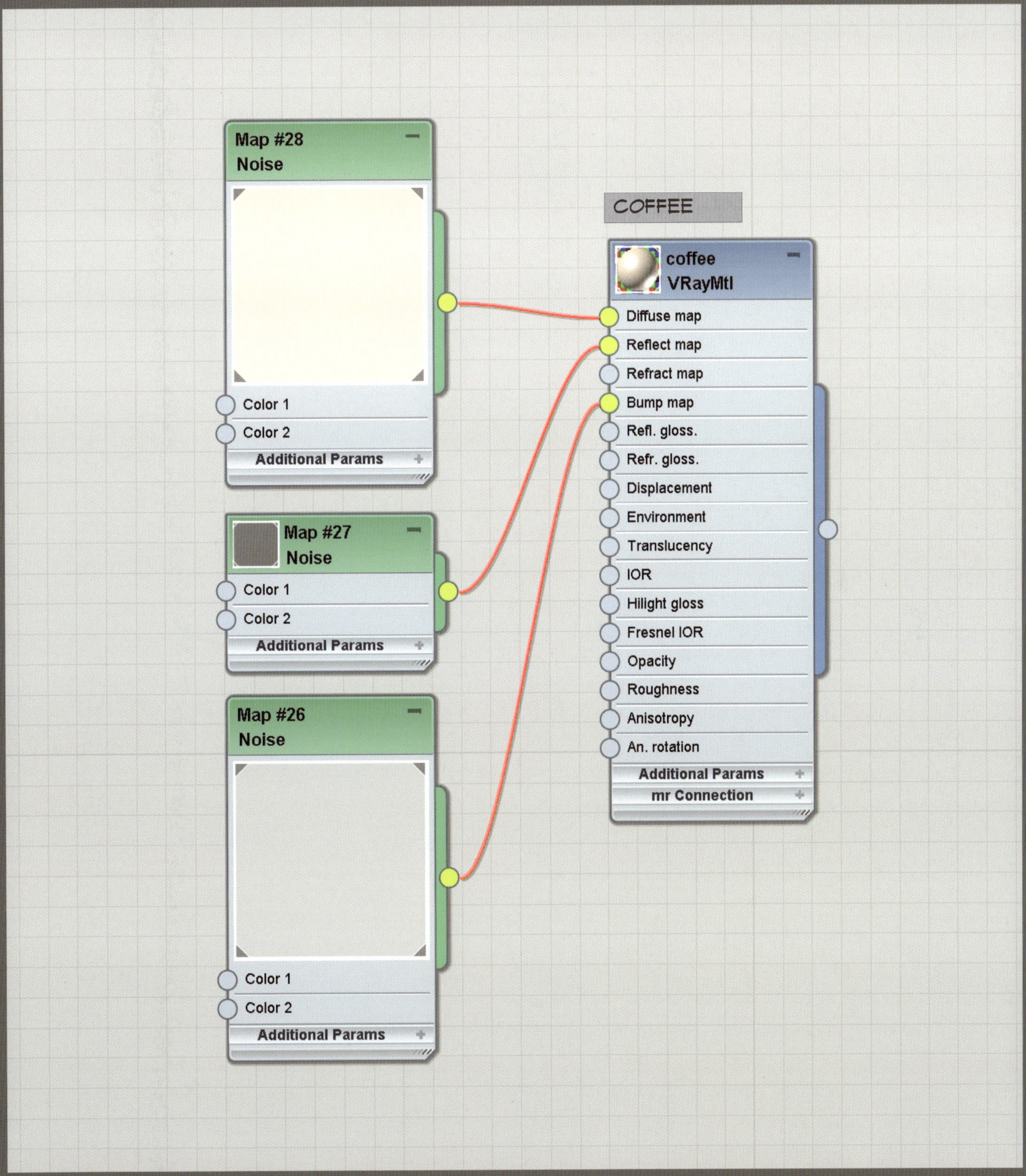

Map #28 — Noise → Diffuse map
Map #27 — Noise → Bump map
Map #26 — Noise → Reflect map

COFFEE — coffee / VRayMtl

- Diffuse map
- Reflect map
- Refract map
- Bump map
- Refl. gloss.
- Refr. gloss.
- Displacement
- Environment
- Translucency
- IOR
- Hilight gloss
- Fresnel IOR
- Opacity
- Roughness
- Anisotropy
- An. rotation
- Additional Params
- mr Connection

TABLE-WARE VOL 1 MADE WITH 3DS MAX AND V-RAY

| TABLE-WARE 018 | 3DS MAX 2010 + FBX (V-RAY) | 7,052 KB | CLASSICAL |

MODIFIERS STACK
- NOT COLLAPSED
- COLLAPSED

TEXTURE
- NOT INCLUDED
- INCLUDED

MAPPING
- UNWRAP
- UVW MAP

TOTAL
- POLYS : 429,761
- VERTS : 331,943

TABLE_WARE

MAX

3DS

FBX - SIMPLE OBJECT WITHOUT MATERIALS (WITH MAPPING AND TEXTURES INCLUDED)

*.MAX - V-RAY 1.5 - OR HIGHER - WITH TEXTURES AND SHADERS

V-RAY - OBJECT PREPARED FOR V-RAY RENDERER (WITH TEXTURES AND SHADERS)

*. MAX 2010 - OR HIGHER

ELEMENT OF MODEL

PARTIAL GROUP

1. TABLE_WARE-A 2. TABLE_WARE-B 3. TABLE_WARE-C 4. TABLE_WARE-D

TABLE-WARE VOL I MADE WITH 3DS MAX AND V-RAY

SLATE MATERIAL EDITOR

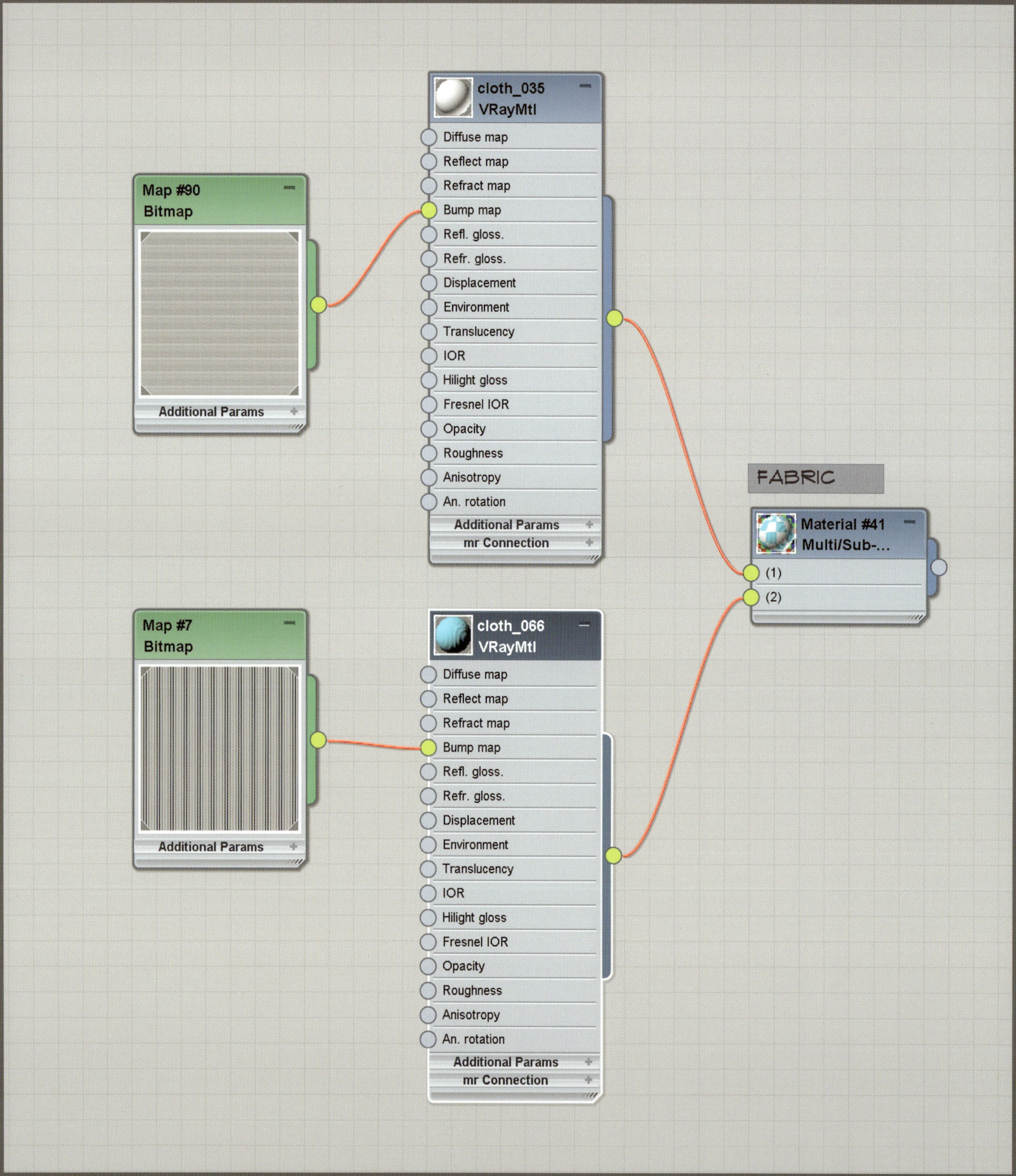

FABRIC

SLATE MATERIAL EDITOR

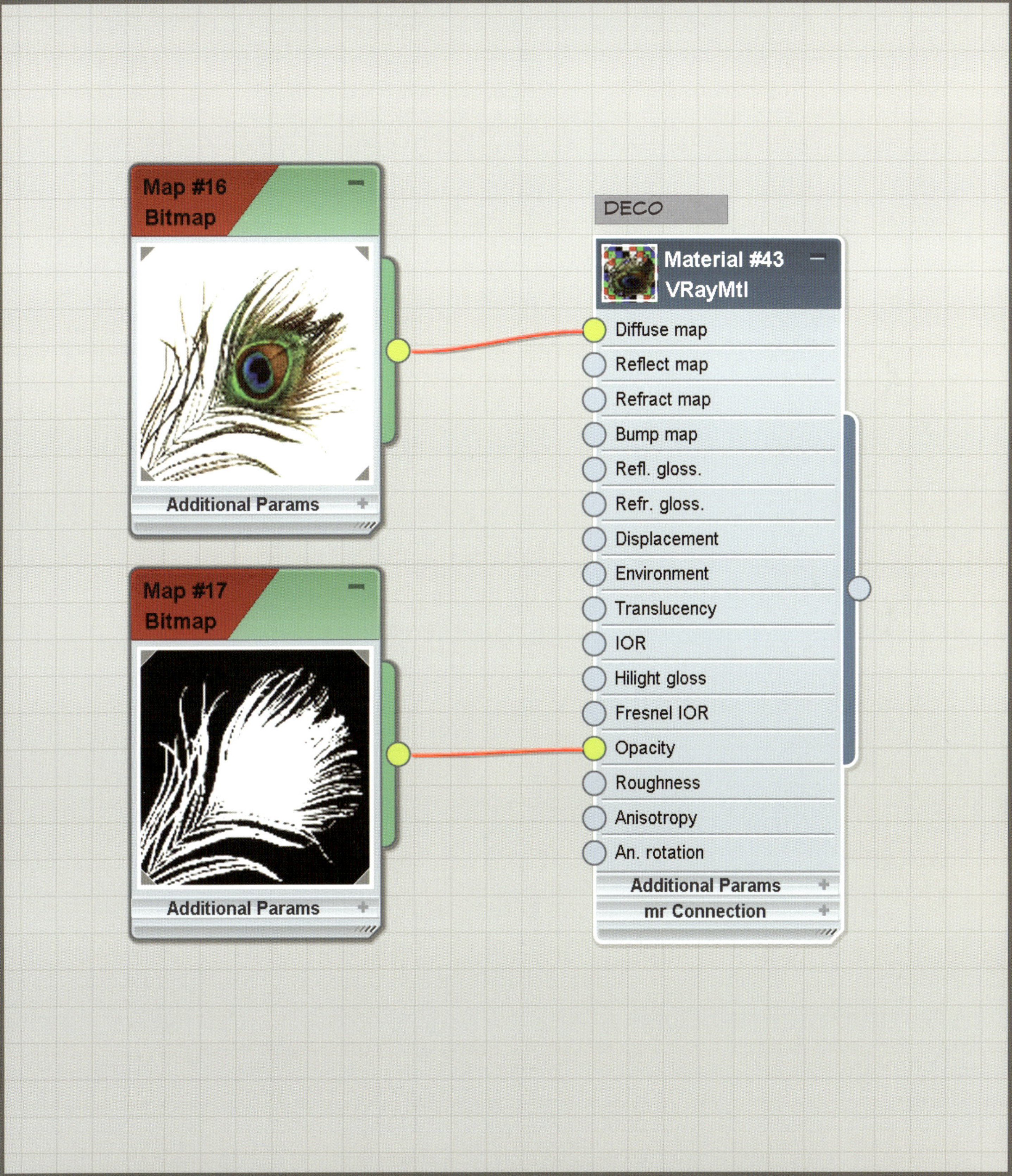

TABLE-WARE 019

3DS MAX 2010 + FBX (V-RAY)   3,504 KB   MODERN

TABLE_WARE

MODIFIERS STACK
- NOT COLLAPSED
- COLLAPSED

TEXTURE
- NOT INCLUDED
- INCLUDED

MAPPING
- UNWRAP
- UVW MAP

TOTAL
- POLYS : 28,122
- VERTS : 27,204

MAX

3DS

FBX – SIMPLE OBJECT WITHOUT MATERIALS (WITH MAPPING AND TEXTURES INCLUDED)

*.MAX – V-RAY 1.5 – OR HIGHER – WITH TEXTURES AND SHADERS

V-RAY – OBJECT PREPARED FOR V-RAY RENDERER (WITH TEXTURES AND SHADERS)

*. MAX 2010 – OR HIGHER

ELEMENT OF MODEL

PARTIAL GROUP

1. TABLE_WARE-A
2. TABLE_WARE-B
3. TABLE_WARE-C

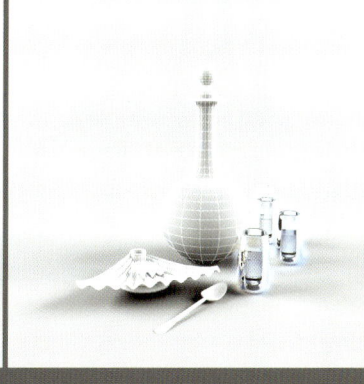

TABLE-WARE VOL 1 MADE WITH 3DS MAX AND V-RAY

THE NEW DIGITAL AGE

SLATE MATERIAL EDITOR

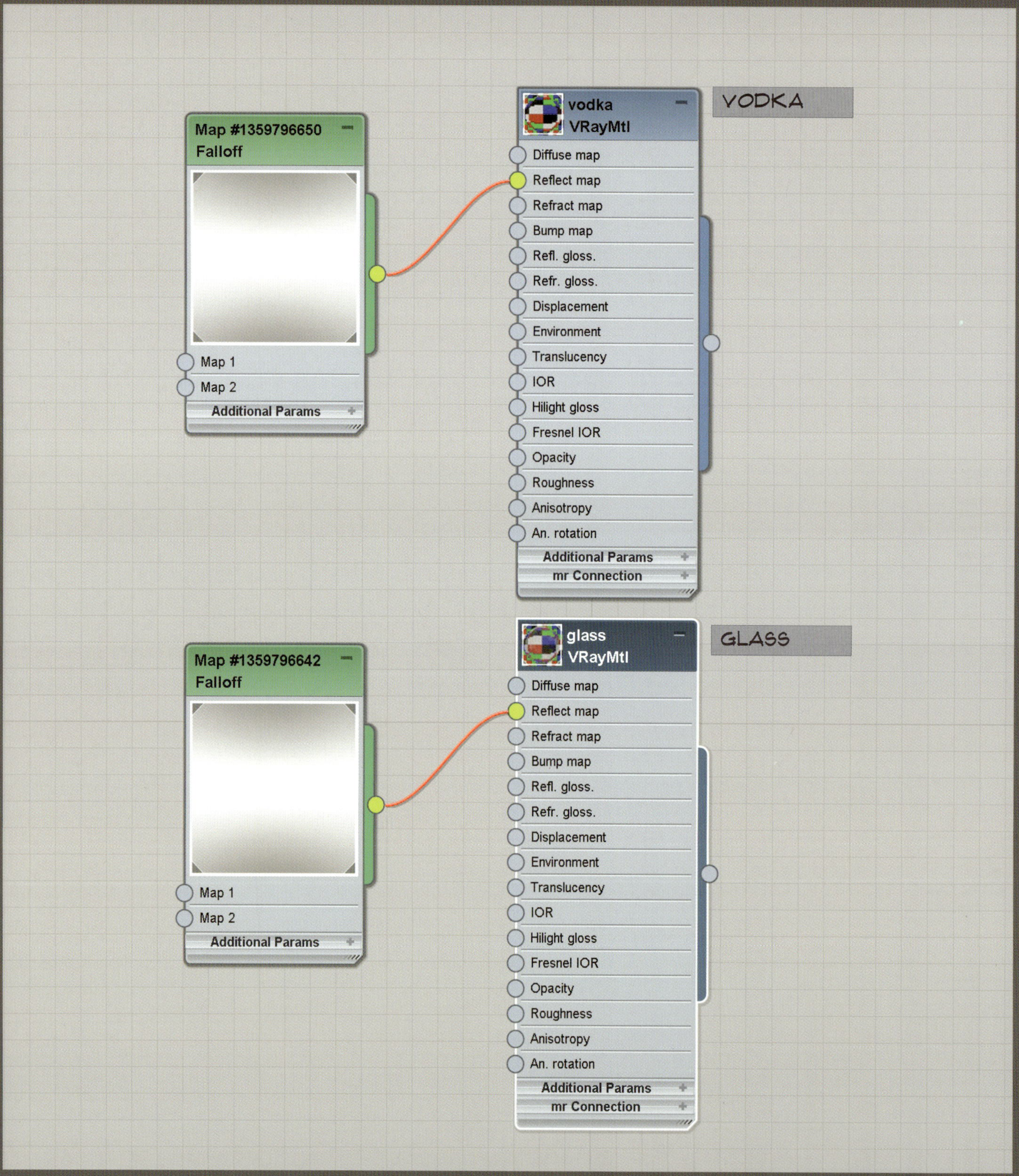

VODKA

GLASS

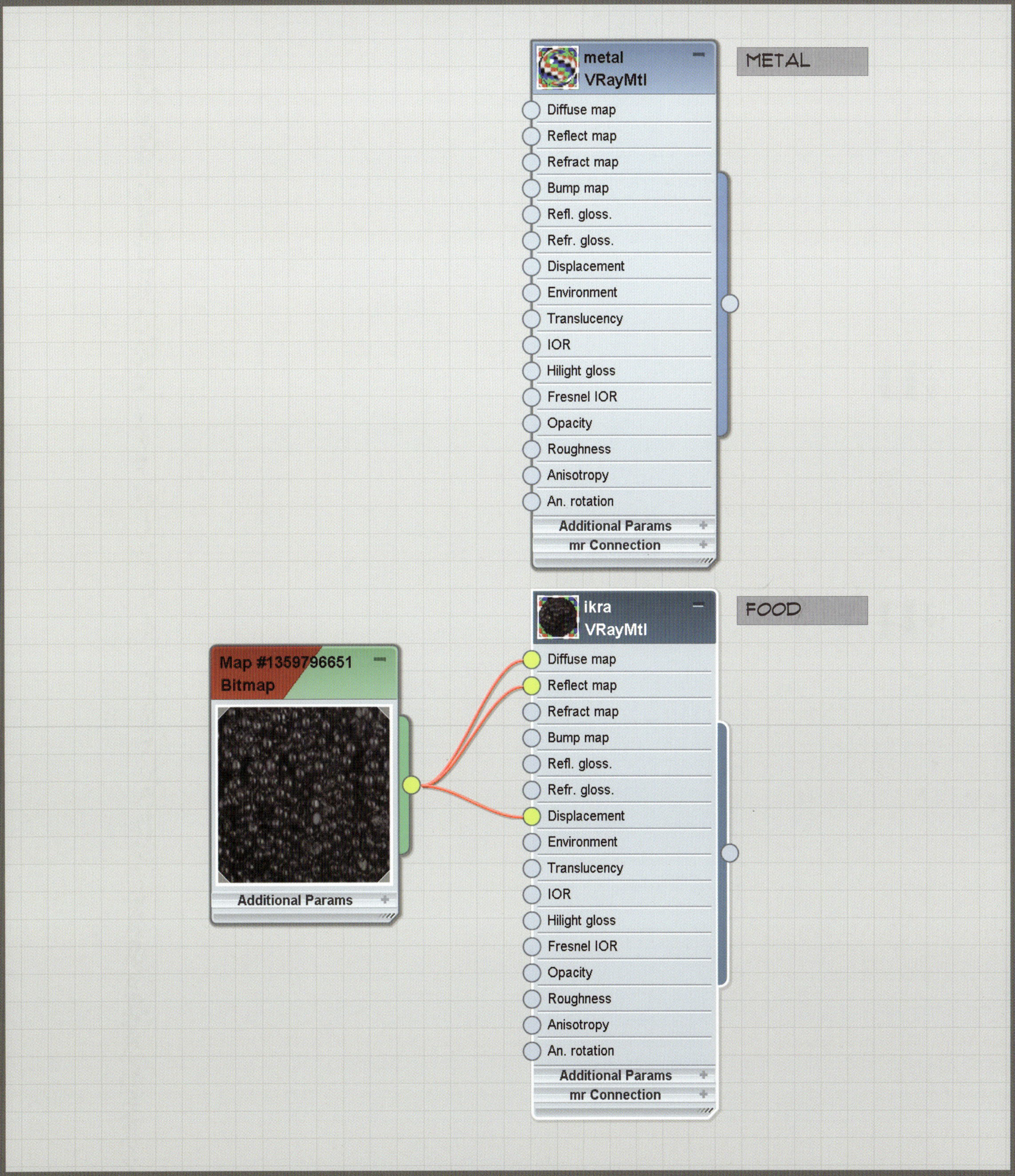

TABLE-WARE 020

⚙ 3DS MAX 2010 + FBX (V-RAY)   🛍 1,244 KB   ☆ MODERN

TABLE_WARE

MODIFIERS STACK
- NOT COLLAPSED
- COLLAPSED

TEXTURE
- NOT INCLUDED
- INCLUDED

MAPPING
- UNWRAP
- UVW MAP

TOTAL
- POLYS : 277,412
- VERTS : 278,589

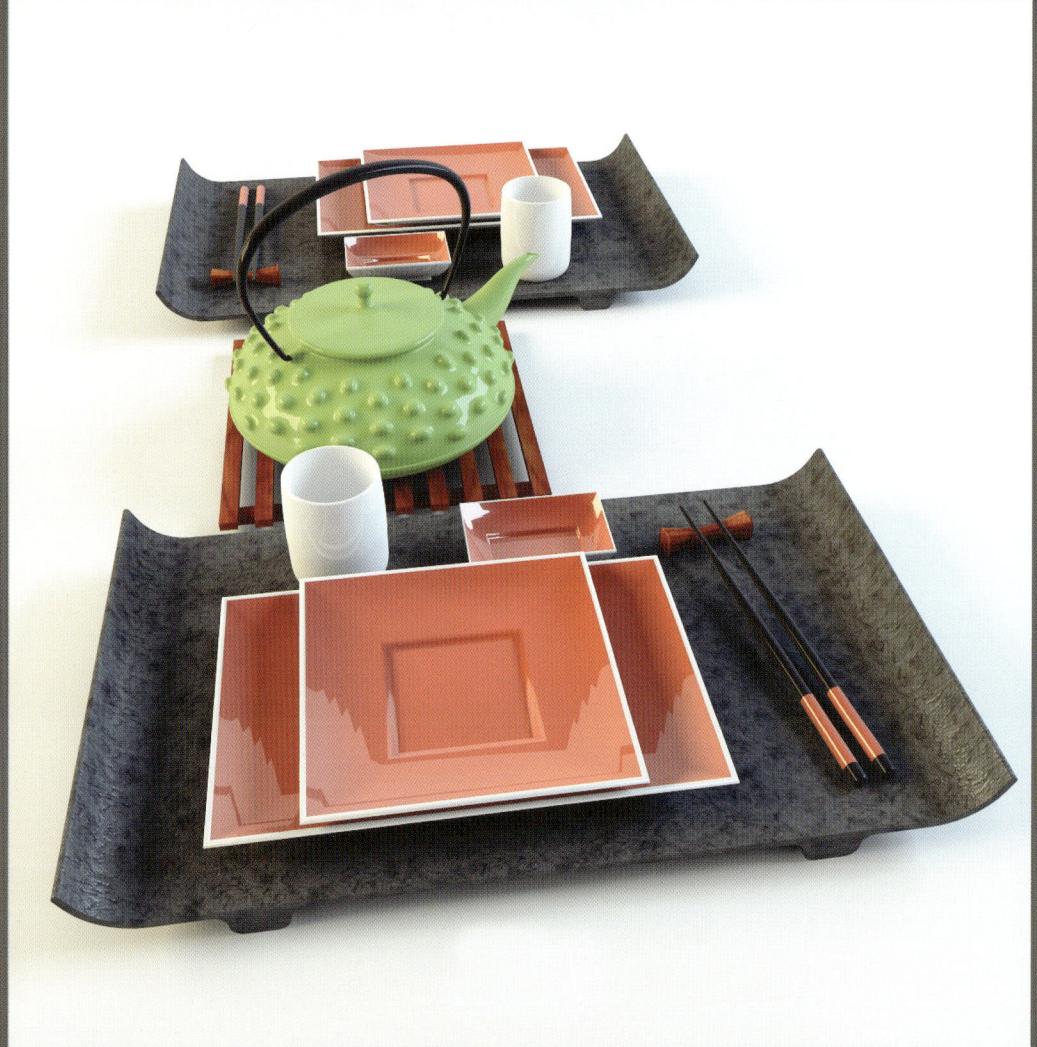

MAX

3DS

FBX – SIMPLE OBJECT WITHOUT MATERIALS (WITH MAPPING AND TEXTURES INCLUDED)

*.MAX – V-RAY 1.5 – OR HIGHER – WITH TEXTURES AND SHADERS

V-RAY – OBJECT PREPARED FOR V-RAY RENDERER (WITH TEXTURES AND SHADERS)

*. MAX 2010 – OR HIGHER

ELEMENT OF MODEL

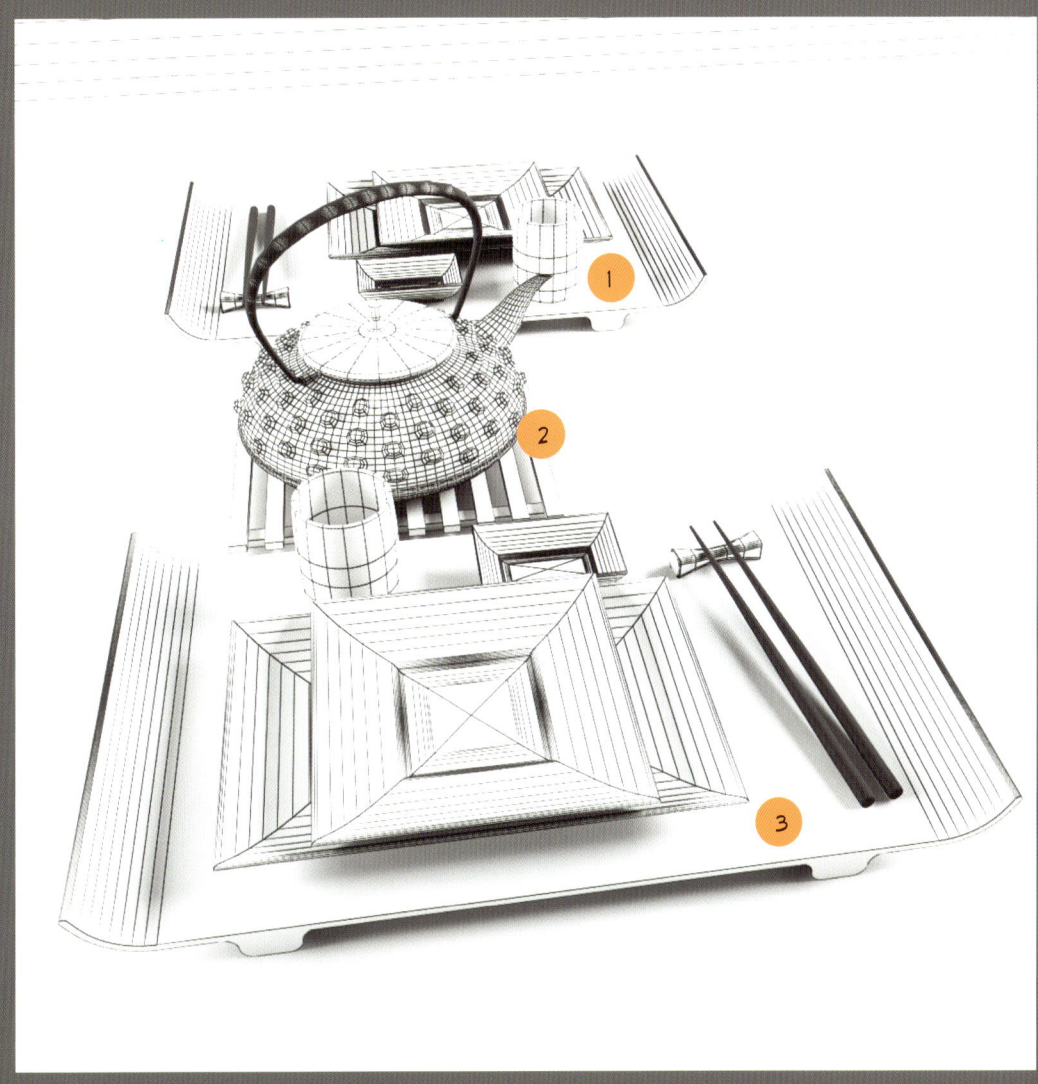

PARTIAL GROUP

1. TABLE_WARE-A   2. TABLE_WARE-B   3. TABLE_WARE-C

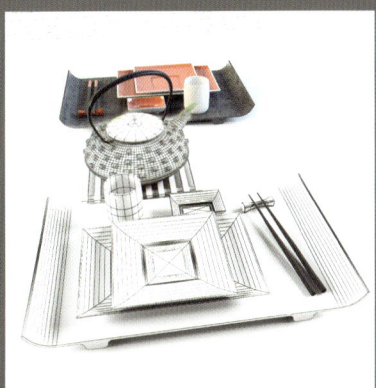

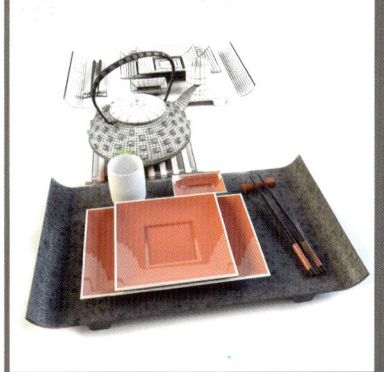

TABLE-WARE VOL 1 MADE WITH 3DS MAX AND V-RAY

SLATE MATERIAL EDITOR

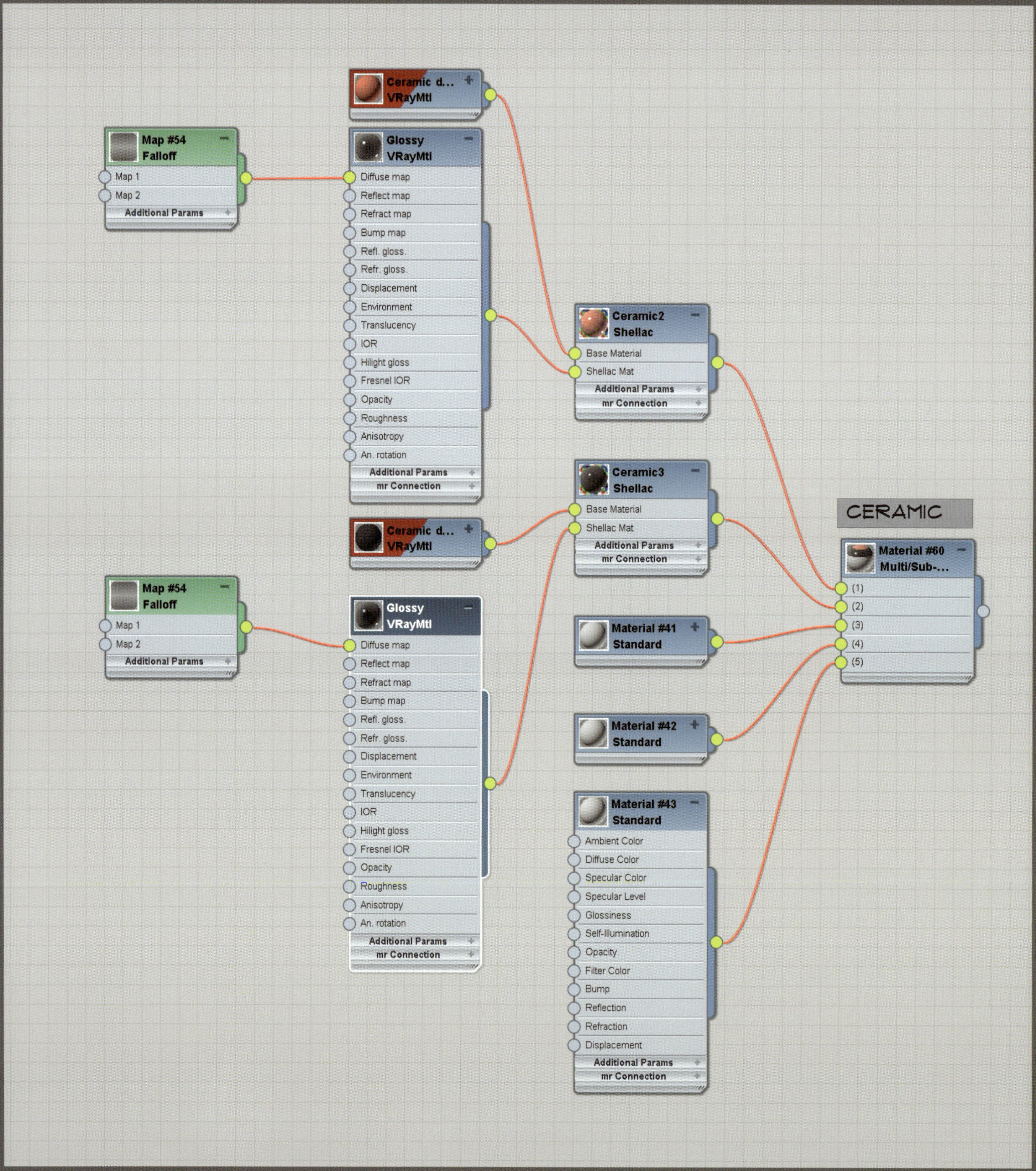

TABLE-WARE VOL 1 MADE WITH 3DS MAX AND V-RAY

SLATE MATERIAL EDITOR

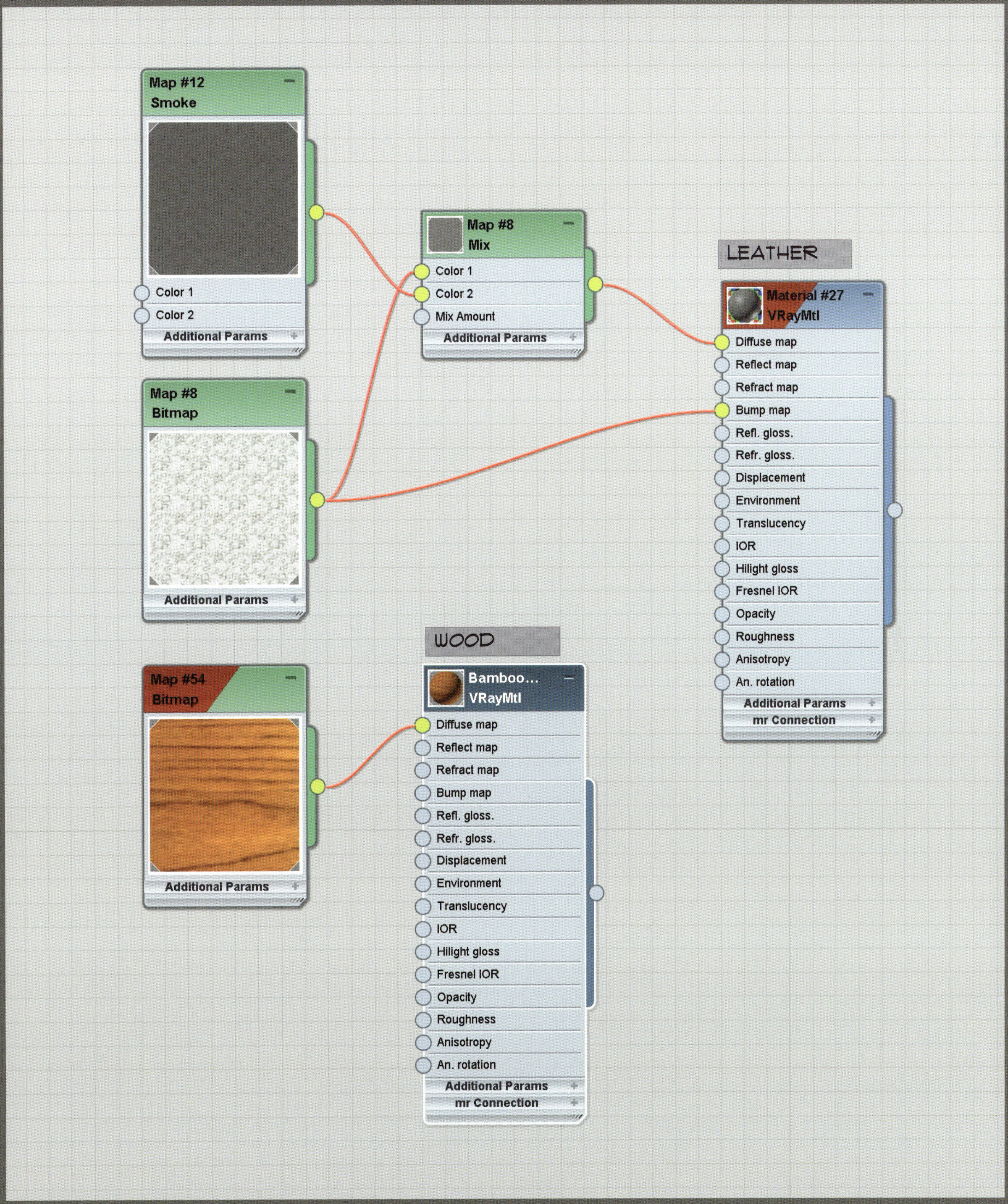

TABLE-WARE VOL I MADE WITH 3DS MAX AND V-RAY

TABLE-WARE 021

MODIFIERS STACK
- NOT COLLAPSED
- COLLAPSED

TEXTURE
- NOT INCLUDED
- INCLUDED

MAPPING
- UNWRAP
- UVW MAP

TOTAL
- POLYS : 184,354
- VERTS : 110,383

3DS MAX 2010 + FBX (V-RAY)   4,304 KB   MODERN

TABLE_WARE

MAX
3DS
FBX – SIMPLE OBJECT WITHOUT MATERIALS (WITH MAPPING AND TEXTURES INCLUDED)
*.MAX – V-RAY 1.5 – OR HIGHER – WITH TEXTURES AND SHADERS
V-RAY – OBJECT PREPARED FOR V-RAY RENDERER (WITH TEXTURES AND SHADERS)

*. MAX 2010 – OR HIGHER

ELEMENT OF MODEL

PARTIAL GROUP

1. TABLE_WARE-A
2. TABLE_WARE-B
3. TABLE_WARE-C
4. TABLE_WARE-D

TABLE-WARE VOL 1 MADE WITH 3DS MAX AND V-RAY

SLATE MATERIAL EDITOR

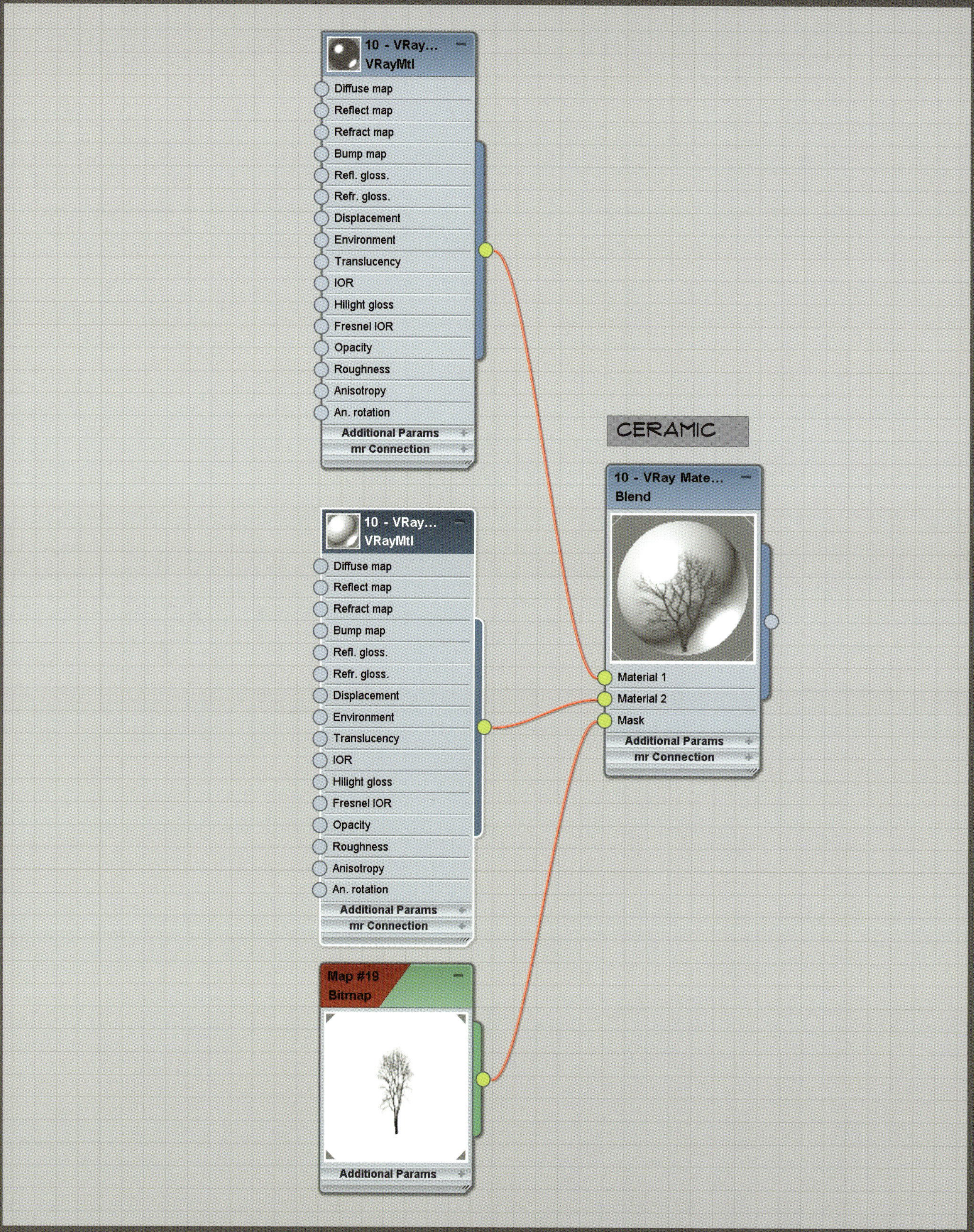

CERAMIC

TABLE-WARE VOL 1 MADE WITH 3DS MAX AND V-RAY

SLATE MATERIAL EDITOR

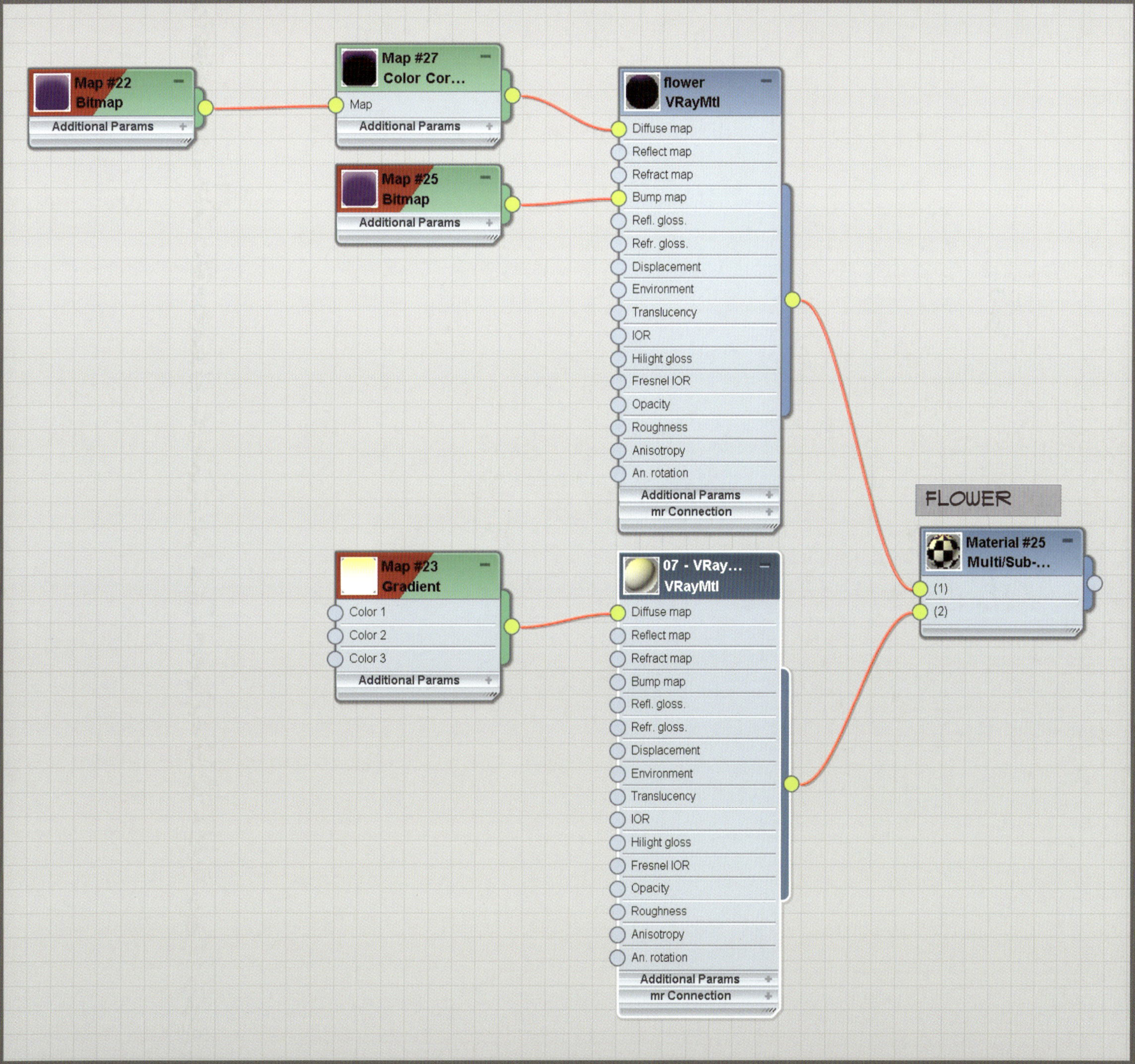

TABLE-WARE VOL 1 MADE WITH 3DS MAX AND V-RAY

| TABLE-WARE 022 | ⚙ 3DS MAX 2010 + FBX (V-RAY) | 52,128 KB | ☆ CLASSICAL |

MODIFIERS STACK
- NOT COLLAPSED
- COLLAPSED

TEXTURE
- NOT INCLUDED
- INCLUDED

MAPPING
- UNWRAP
- UVW MAP

TOTAL
- POLYS : 355,671
- VERTS : 361,969

TABLE_WARE

MAX

3DS

FBX – SIMPLE OBJECT WITHOUT MATERIALS (WITH MAPPING AND TEXTURES INCLUDED)

*.MAX – V-RAY 1.5 – OR HIGHER – WITH TEXTURES AND SHADERS

V-RAY – OBJECT PREPARED FOR V-RAY RENDERER (WITH TEXTURES AND SHADERS)

*. MAX 2010 – OR HIGHER

ELEMENT OF MODEL

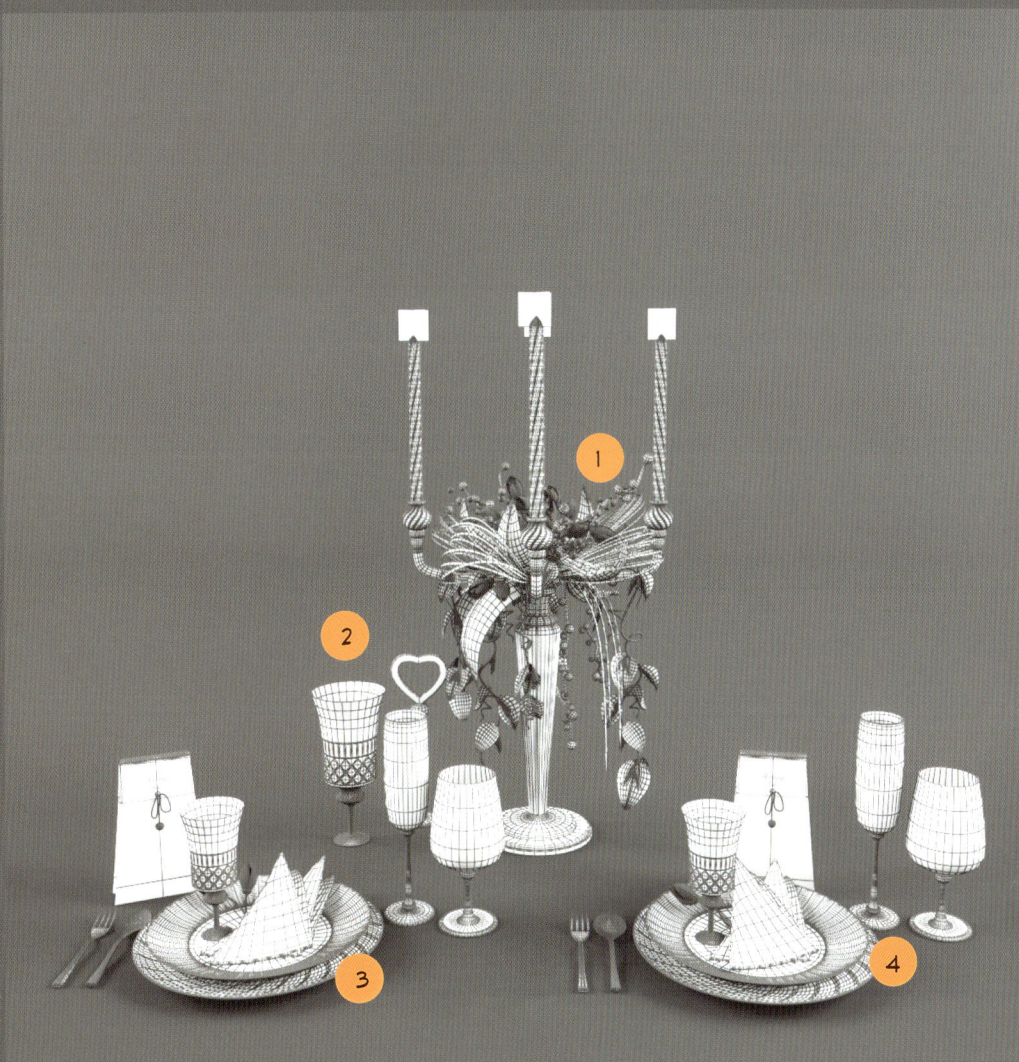

PARTIAL GROUP

1. TABLE_WARE-A   2. TABLE_WARE-B   3. TABLE_WARE-C   4. TABLE_WARE-D

TABLE-WARE VOL I MADE WITH 3DS MAX AND V-RAY

SLATE MATERIAL EDITOR

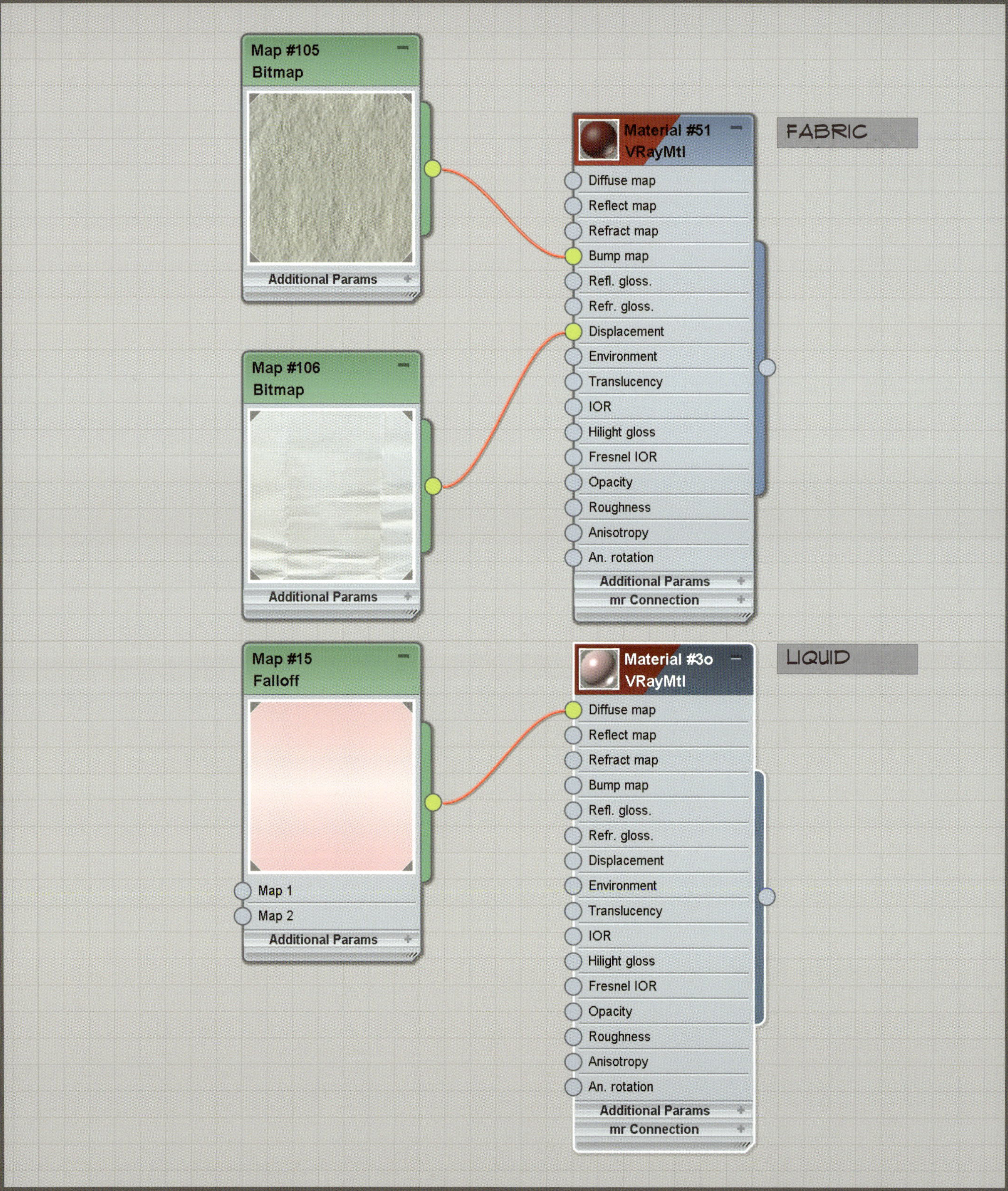

TABLE-WARE VOL I MADE WITH 3DS MAX AND V-RAY

MADE WITH: THE NEW DIGITAL AGE

SLATE MATERIAL EDITOR

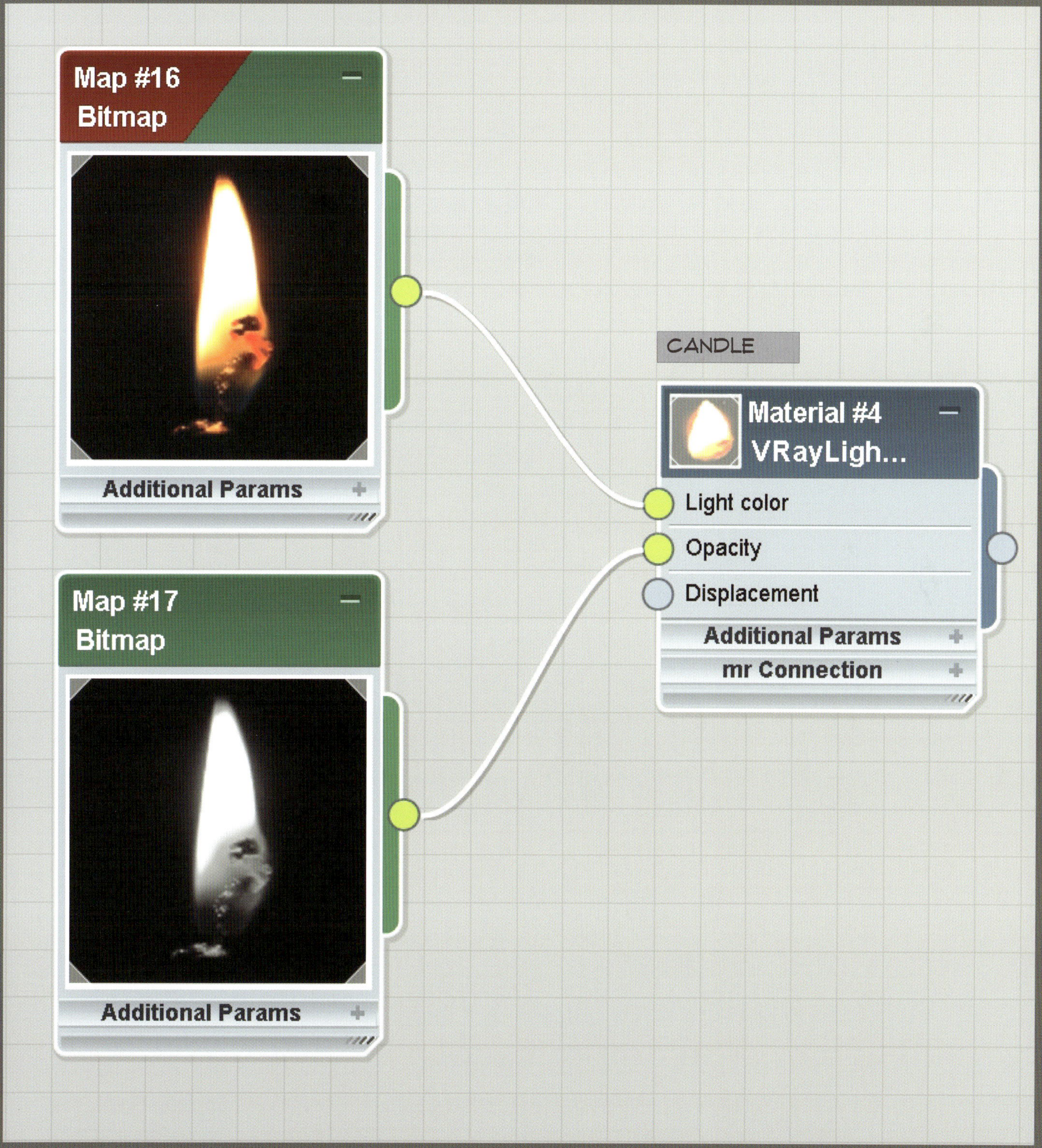

TABLE-WARE VOL 1 MADE WITH 3DS MAX AND V-RAY

| TABLE-WARE 023 | 3DS MAX 2010 + FBX (V-RAY) | 16,752 KB | CLASSICAL |

TABLE_WARE

MODIFIERS STACK
- NOT COLLAPSED
- COLLAPSED

TEXTURE
- NOT INCLUDED
- INCLUDED

MAPPING
- UNWRAP
- UVW MAP

TOTAL
- POLYS : 233,214
- VERTS : 177,161

MAX

3DS

FBX - SIMPLE OBJECT WITHOUT MATERIALS (WITH MAPPING AND TEXTURES INCLUDED)

*.MAX - V-RAY 1.5 - OR HIGHER - WITH TEXTURES AND SHADERS

V-RAY - OBJECT PREPARED FOR V-RAY RENDERER (WITH TEXTURES AND SHADERS)

*. MAX 2010 - OR HIGHER

MADE WITH:
THE NEW DIGITAL AGE

ELEMENT OF MODEL

PARTIAL GROUP

1. TABLE_WARE-A 2. TABLE_WARE-B 3. TABLE_WARE-C 4. TABLE_WARE-D

TABLE-WARE VOL 1 MADE WITH 3DS MAX AND V-RAY

SLATE MATERIAL EDITOR

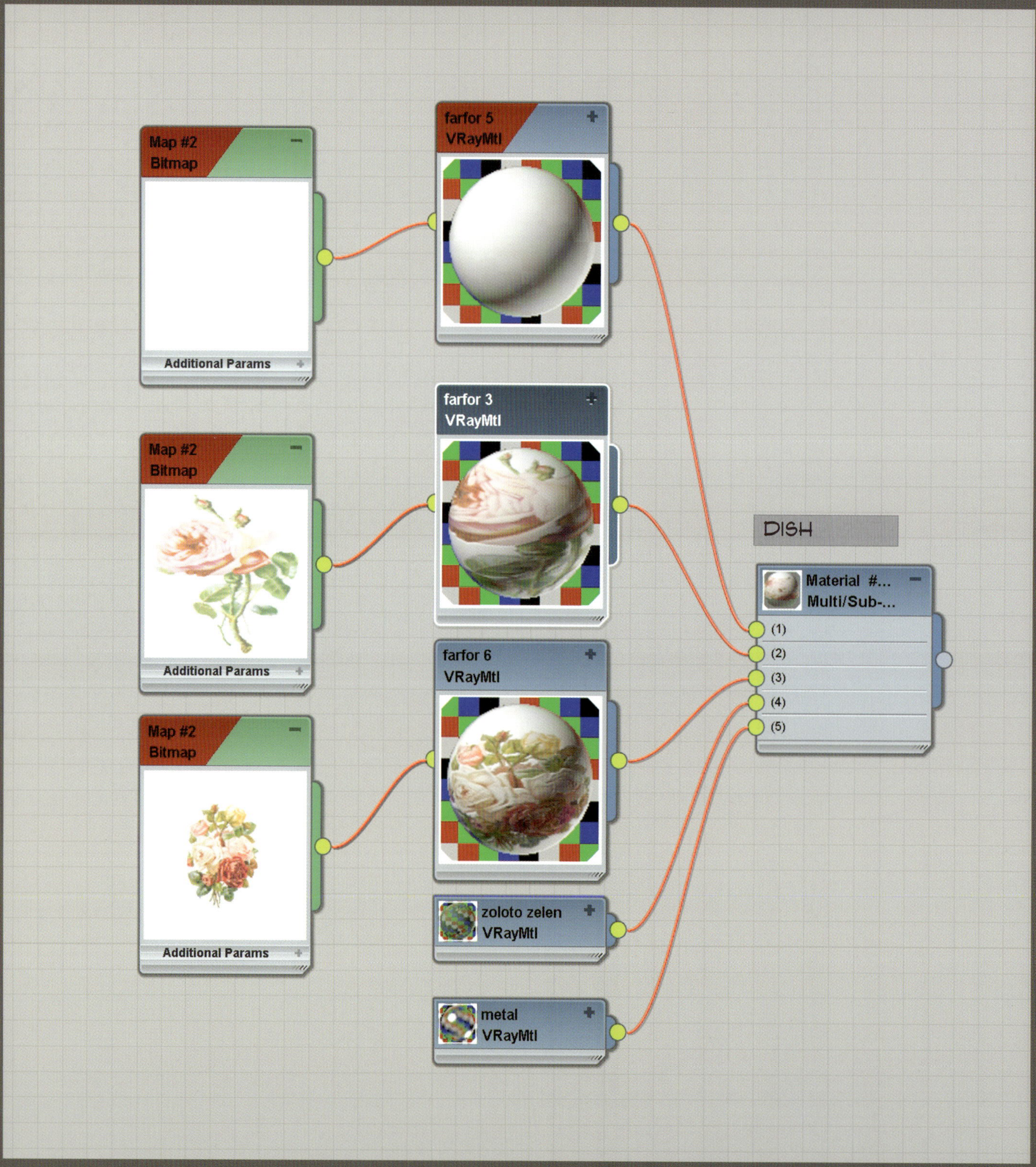

MADE WITH:
THE NEW DIGITAL AGE

SLATE MATERIAL EDITOR

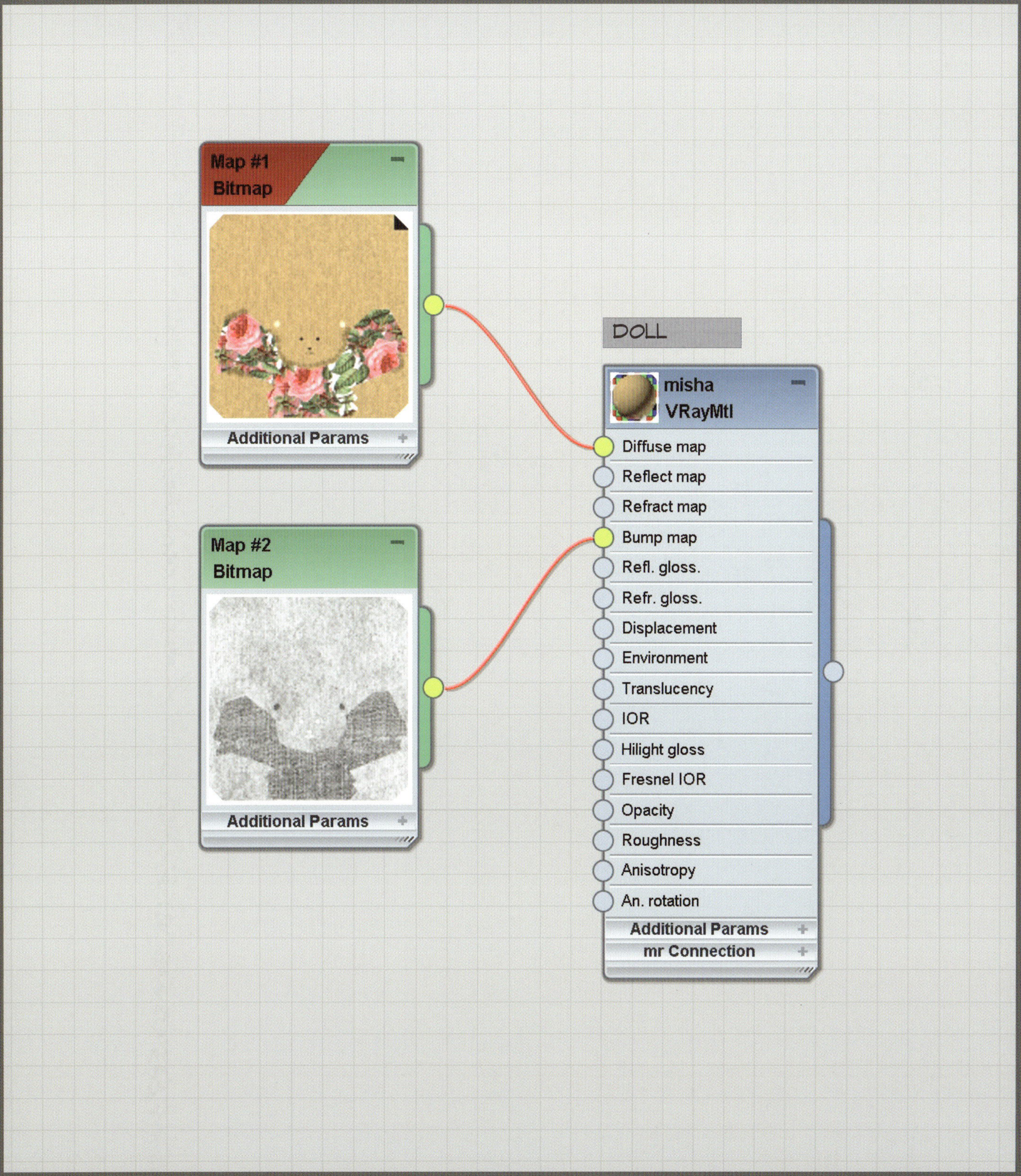

Map #1 Bitmap
Additional Params

Map #2 Bitmap
Additional Params

DOLL

misha VRayMtl
- Diffuse map
- Reflect map
- Refract map
- Bump map
- Refl. gloss.
- Refr. gloss.
- Displacement
- Environment
- Translucency
- IOR
- Hilight gloss
- Fresnel IOR
- Opacity
- Roughness
- Anisotropy
- An. rotation

Additional Params
mr Connection

TABLE-WARE VOL 1 MADE WITH 3DS MAX AND V-RAY

TABLE-WARE 024

3DS MAX 2010 + FBX (V-RAY)   5,316 KB   CLASSICAL

MODIFIERS STACK
NOT COLLAPSED
COLLAPSED

TEXTURE
NOT INCLUDED
INCLUDED

MAPPING
UNWRAP
UVW MAP

TOTAL
POLYS : 35,600
VERTS : 35,062

TABLE_WARE

MAX

3DS

FBX – SIMPLE OBJECT WITHOUT MATERIALS (WITH MAPPING AND TEXTURES INCLUDED)

*.MAX – V-RAY 1.5 – OR HIGHER – WITH TEXTURES AND SHADERS

V-RAY – OBJECT PREPARED FOR V-RAY RENDERER (WITH TEXTURES AND SHADERS)

*. MAX 2010 – OR HIGHER

ELEMENT OF MODEL

PARTIAL GROUP

1. TABLE_WARE-A   2. TABLE_WARE-B   3. TABLE_WARE-C   4. TABLE_WARE-D

TABLE-WARE VOL 1 MADE WITH 3DS MAX AND V-RAY

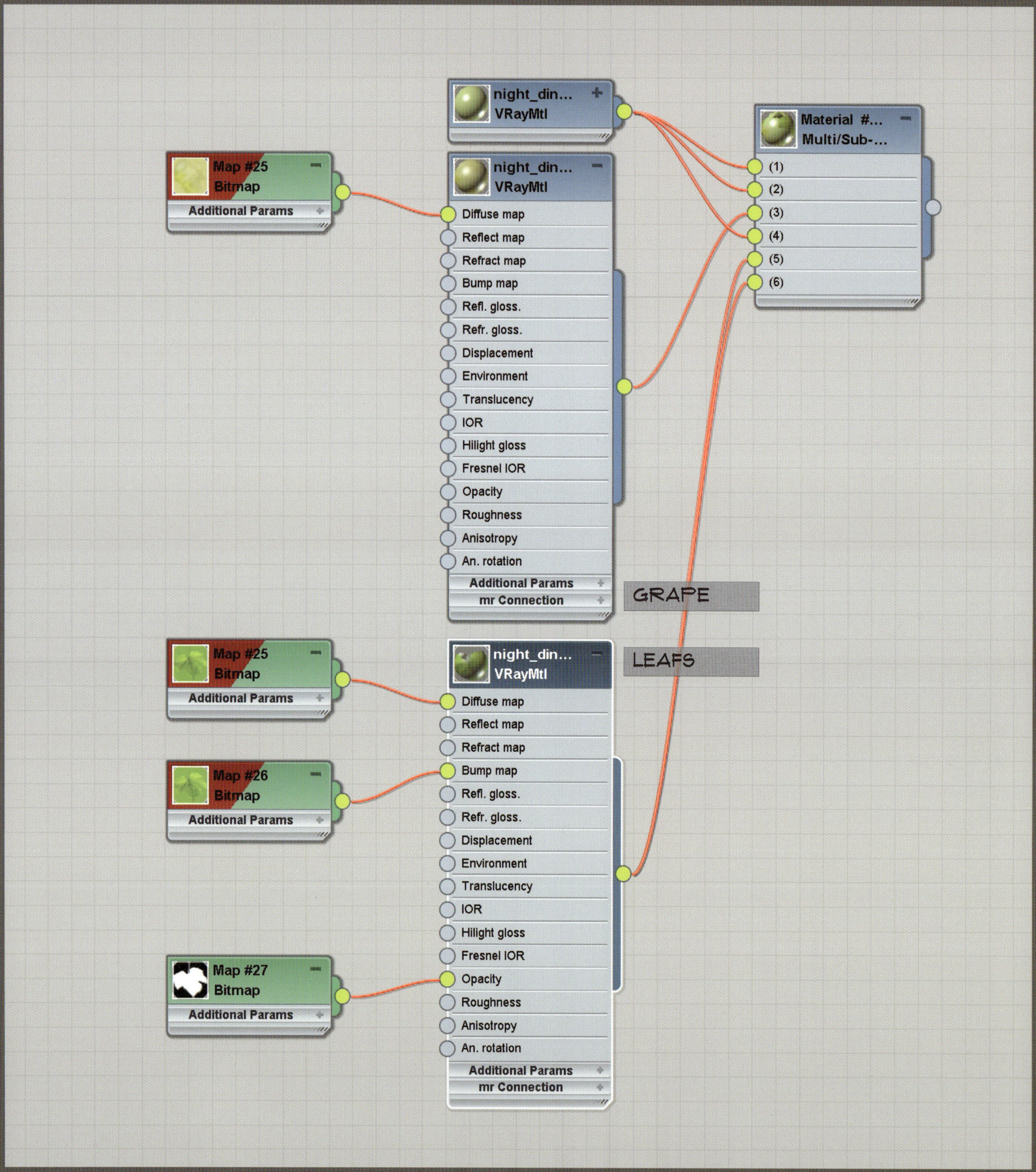

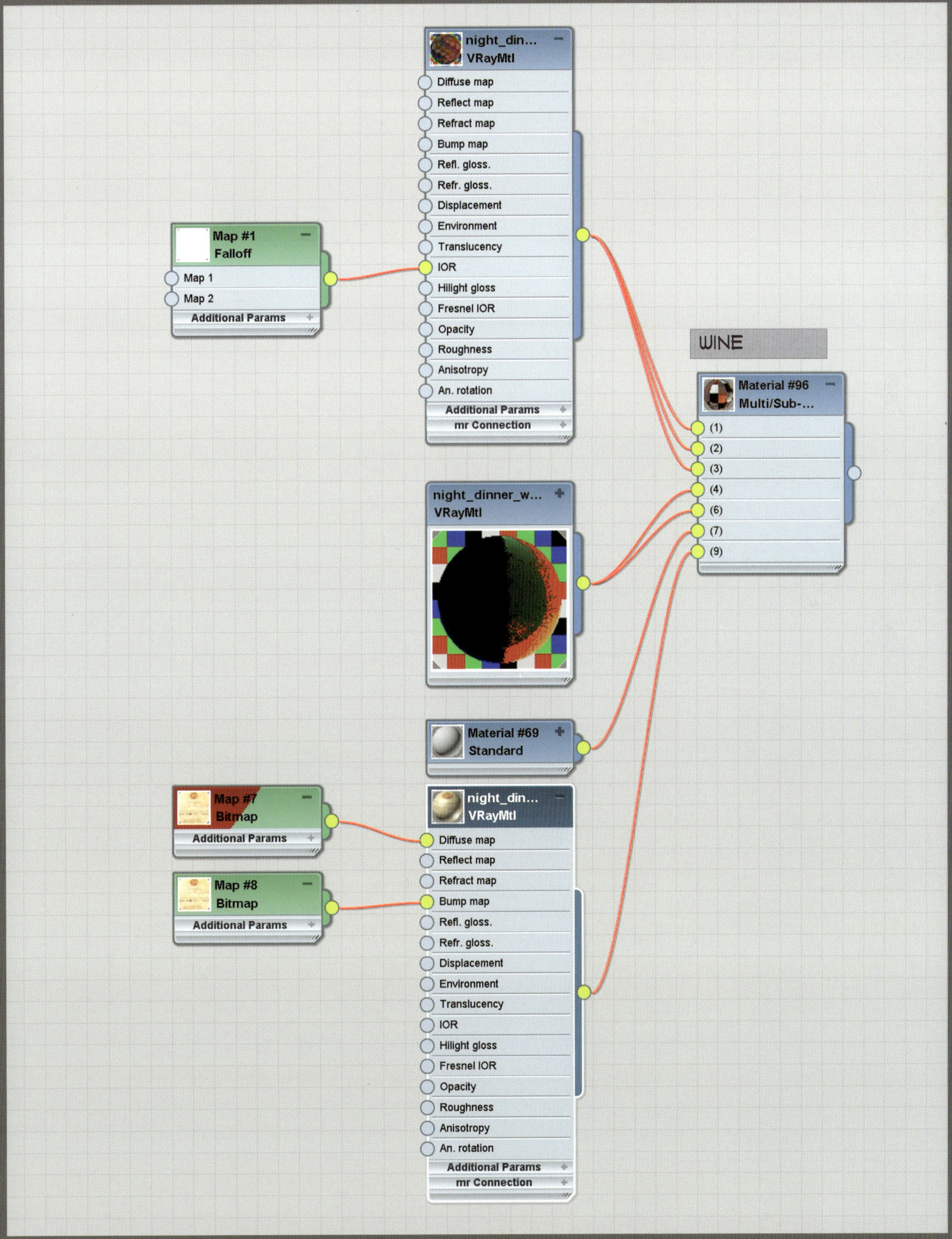

TABLE-WARE 025

MODIFIERS STACK
- NOT COLLAPSED
- COLLAPSED

TEXTURE
- NOT INCLUDED
- INCLUDED

MAPPING
- UNWRAP
- UVW MAP

TOTAL
- POLYS : 362,510
- VERTS : 351,268

3DS MAX 2010 + FBX (V-RAY)   19,980 KB   CLASSICAL

TABLE_WARE

MAX

3DS

FBX - SIMPLE OBJECT WITHOUT MATERIALS (WITH MAPPING AND TEXTURES INCLUDED)

*.MAX - V-RAY 1.5 - OR HIGHER - WITH TEXTURES AND SHADERS

V-RAY - OBJECT PREPARED FOR V-RAY RENDERER (WITH TEXTURES AND SHADERS)

*. MAX 2010 - OR HIGHER

ELEMENT OF MODEL

PARTIAL GROUP

1. TABLE_WARE-A   2. TABLE_WARE-B   3. TABLE_WARE-C   4. TABLE_WARE-D

TABLE-WARE VOL 1 MADE WITH 3DS MAX AND V-RAY

SLATE MATERIAL EDITOR

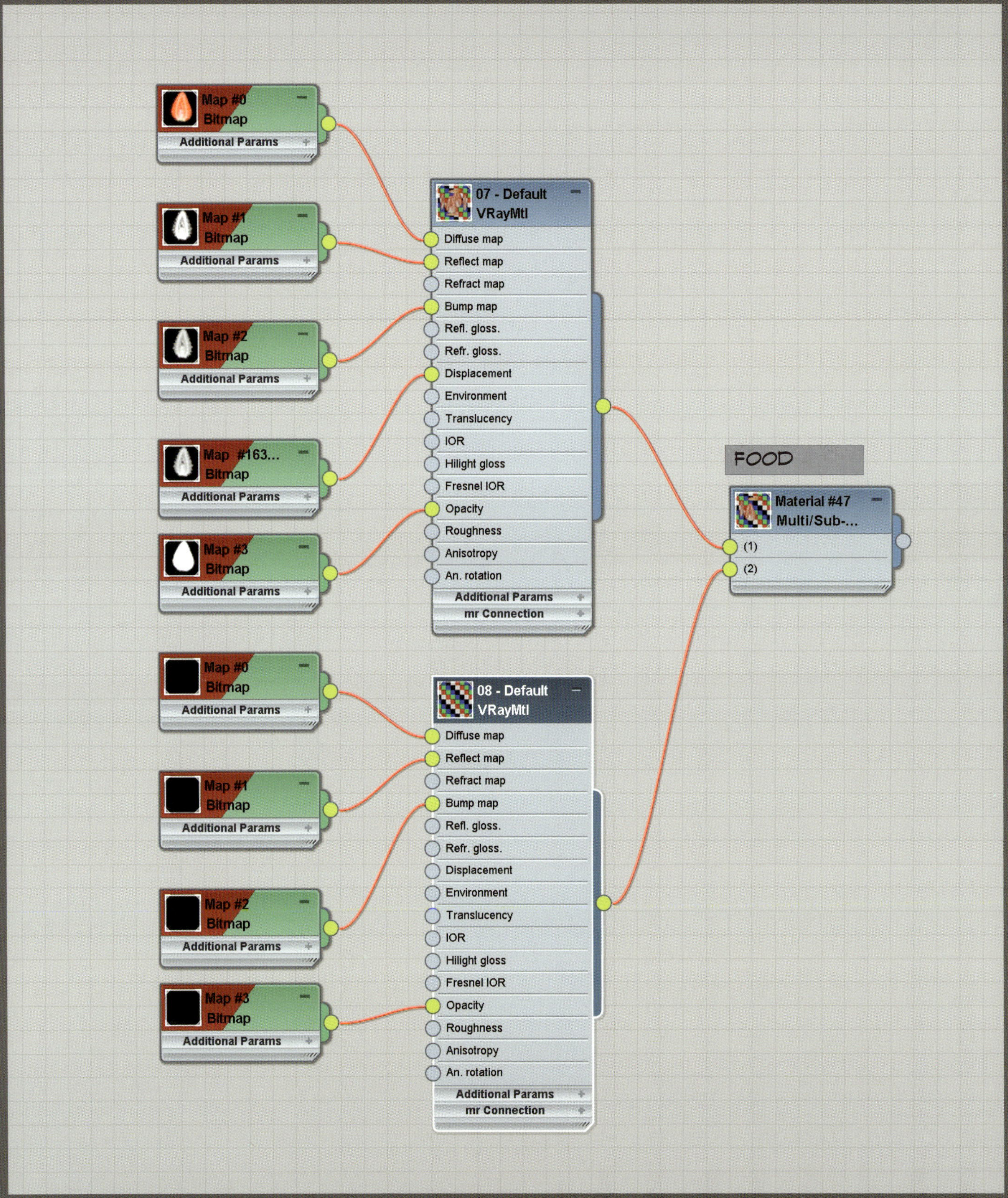

SLATE MATERIAL EDITOR

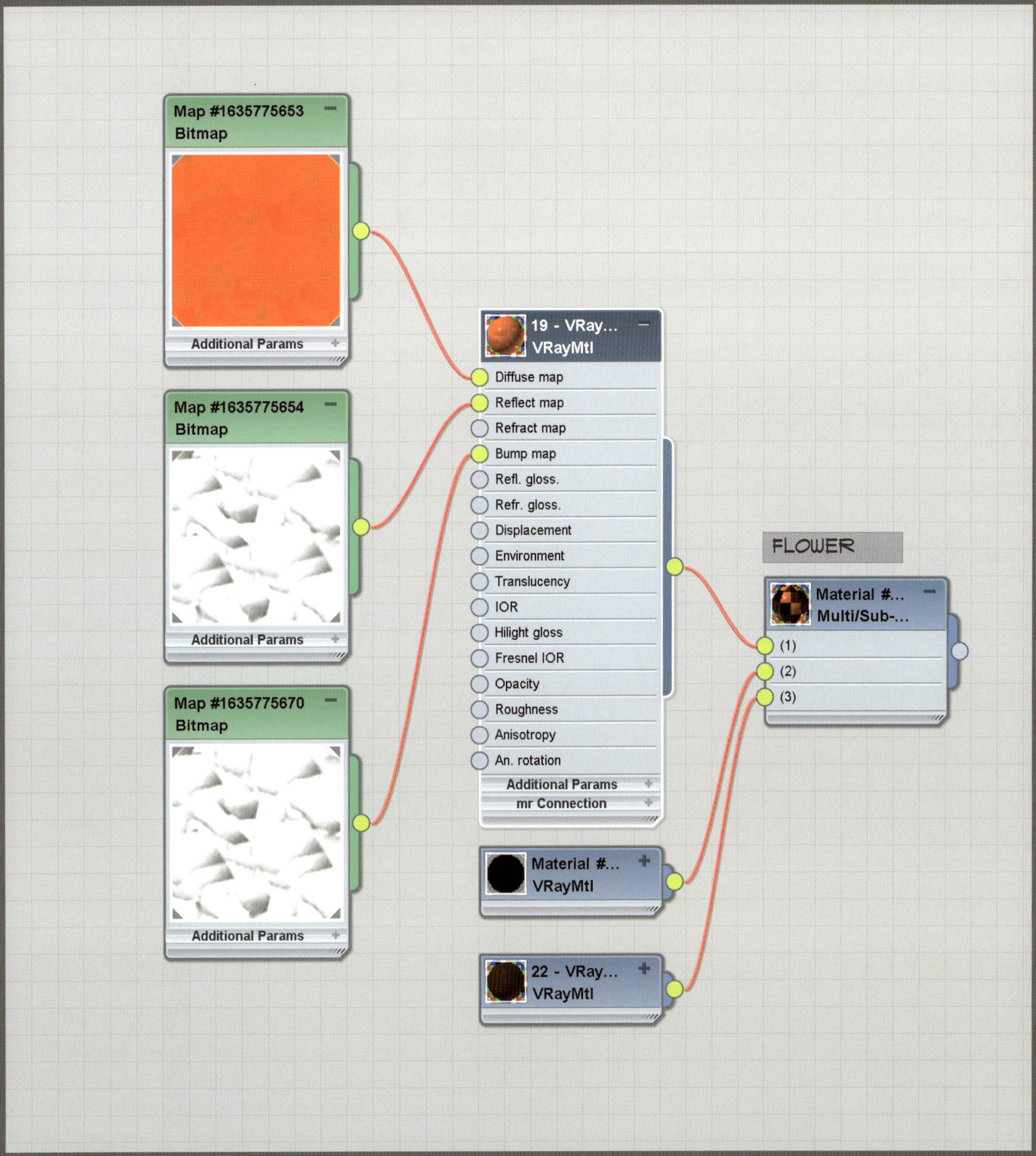

FLOWER

TABLE-WARE VOL I | MADE WITH 3DS MAX AND V-RAY

TABLE-WARE 026

3DS MAX 2010 + FBX (V-RAY)   10,104 KB   MODERN

MODIFIERS STACK
- NOT COLLAPSED
- COLLAPSED

TEXTURE
- NOT INCLUDED
- INCLUDED

MAPPING
- UNWRAP
- UVW MAP

TOTAL
- POLYS : 74,181
- VERTS : 74,608

TABLE_WARE

MAX
3DS
FBX – SIMPLE OBJECT WITHOUT MATERIALS (WITH MAPPING AND TEXTURES INCLUDED)
*.MAX – V-RAY 1.5 – OR HIGHER – WITH TEXTURES AND SHADERS
V-RAY – OBJECT PREPARED FOR V-RAY RENDERER (WITH TEXTURES AND SHADERS)

*. MAX 2010 – OR HIGHER

ELEMENT OF MODEL

PARTIAL GROUP

1. TABLE_WARE-A   2. TABLE_WARE-B   3. TABLE_WARE-C   4. TABLE_WARE-D

TABLE-WARE VOL I MADE WITH 3DS MAX AND V-RAY

SLATE MATERIAL EDITOR

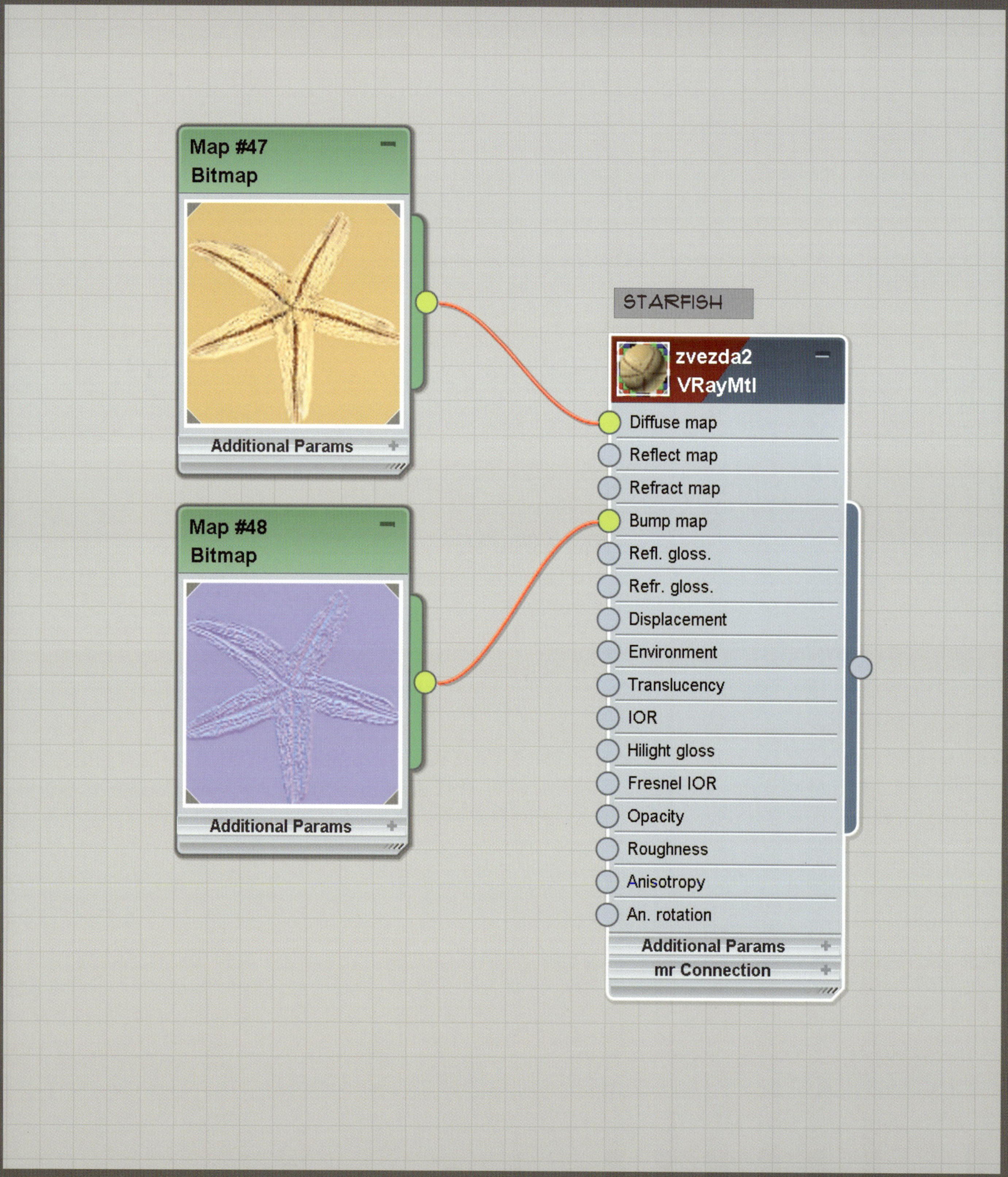

SLATE MATERIAL EDITOR

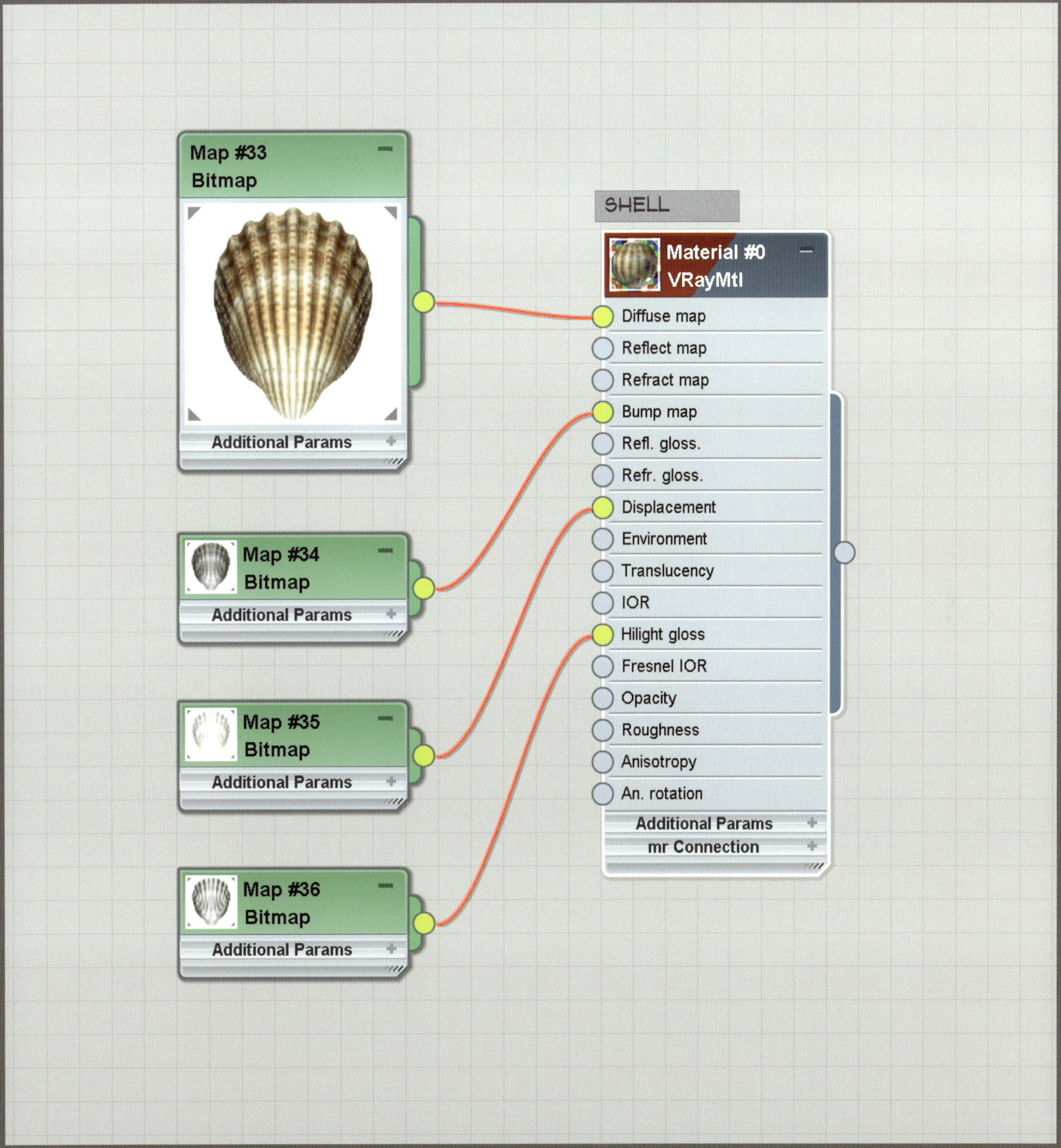

TABLE-WARE VOL I MADE WITH 3DS MAX AND V-RAY

| TABLE-WARE 027 | 3DS MAX 2010 + FBX (V-RAY) | 12,072 KB | CLASSICAL |

MODIFIERS STACK
- NOT COLLAPSED
- COLLAPSED

TEXTURE
- NOT INCLUDED
- INCLUDED

MAPPING
- UNWRAP
- UVW MAP

TOTAL
- POLYS : 1,827,154
- VERTS : 974,700

TABLE_WARE

MAX
3DS
FBX – SIMPLE OBJECT WITHOUT MATERIALS (WITH MAPPING AND TEXTURES INCLUDED)
*.MAX – V-RAY 1.5 – OR HIGHER – WITH TEXTURES AND SHADERS
V-RAY – OBJECT PREPARED FOR V-RAY RENDERER (WITH TEXTURES AND SHADERS)

*. MAX 2010 – OR HIGHER

ELEMENT OF MODEL

PARTIAL GROUP

1. TABLE_WARE-A
2. TABLE_WARE-B
3. TABLE_WARE-C
4. TABLE_WARE-D

TABLE-WARE VOL I MADE WITH 3DS MAX AND V-RAY

SLATE MATERIAL EDITOR

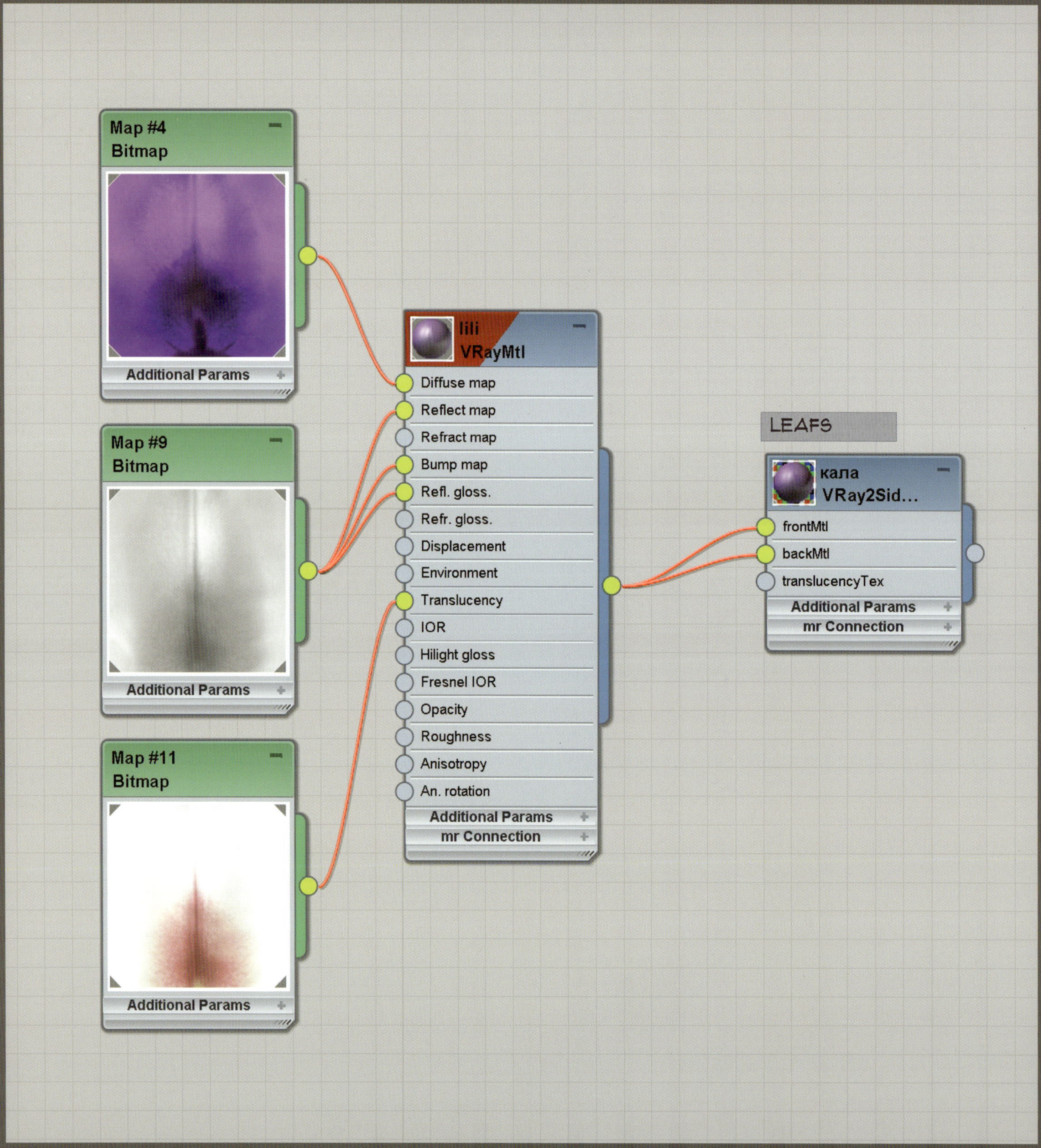

| TABLE-WARE 028 | 3DS MAX 2010 + FBX (V-RAY) | 27,308 KB | CLASSICAL |

MODIFIERS STACK
- NOT COLLAPSED
- COLLAPSED

TEXTURE
- NOT INCLUDED
- INCLUDED

MAPPING
- UNWRAP
- UVW MAP

TOTAL
- POLYS : 286,359
- VERTS : 210,069

TABLE_WARE

MAX
3DS
FBX - SIMPLE OBJECT WITHOUT MATERIALS (WITH MAPPING AND TEXTURES INCLUDED)
*.MAX - V-RAY 1.5 - OR HIGHER - WITH TEXTURES AND SHADERS
V-RAY - OBJECT PREPARED FOR V-RAY RENDERER (WITH TEXTURES AND SHADERS)

*. MAX 2010 - OR HIGHER

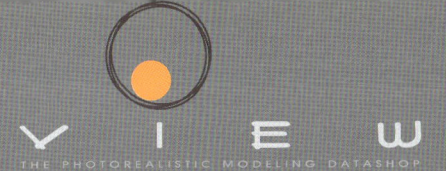

ELEMENT OF MODEL

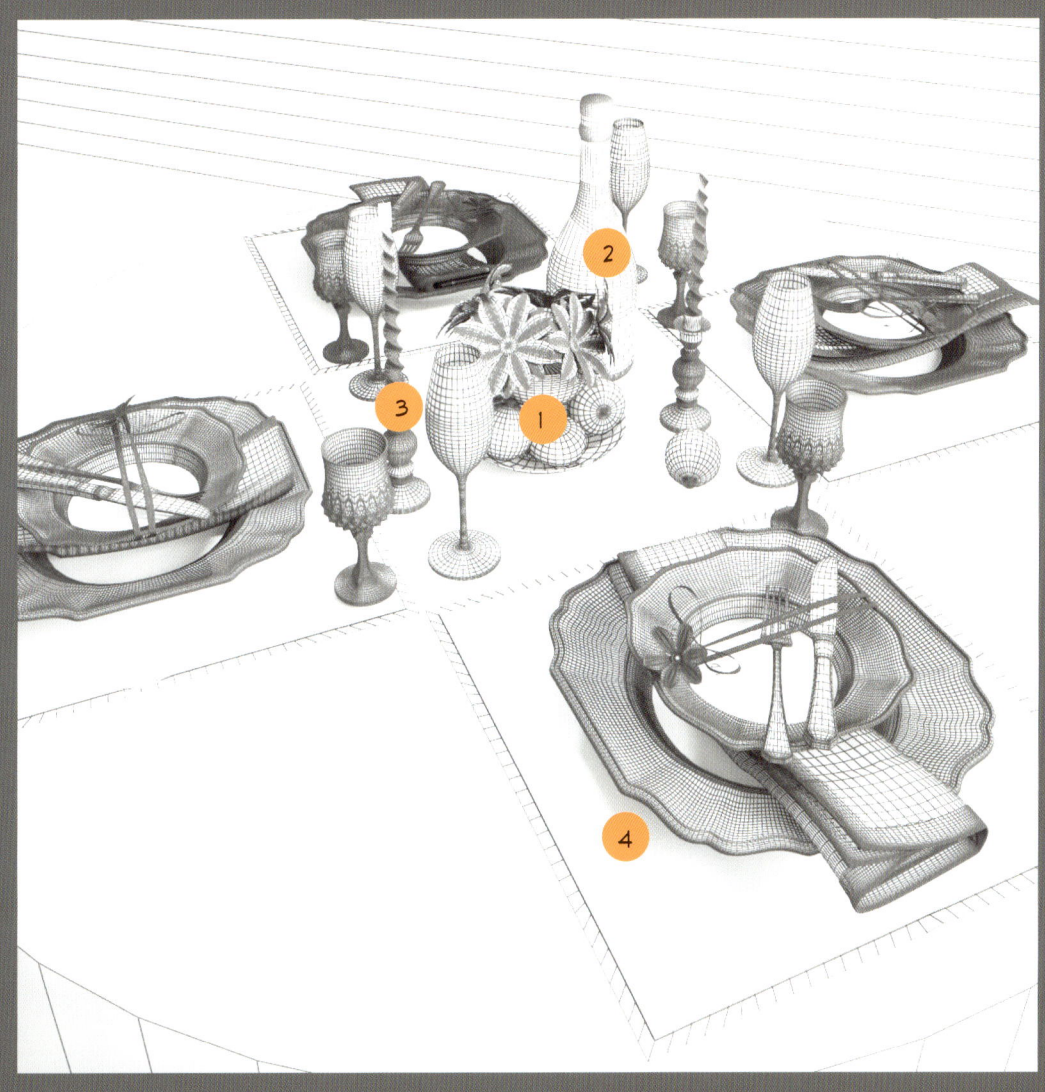

PARTIAL GROUP

1. TABLE_WARE-A
2. TABLE_WARE-B
3. TABLE_WARE-C
4. TABLE_WARE-D

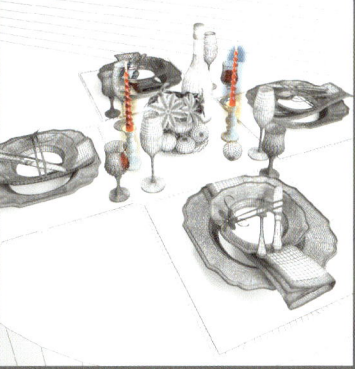

TABLE-WARE VOL 1 MADE WITH 3DS MAX AND V-RAY

SLATE MATERIAL EDITOR

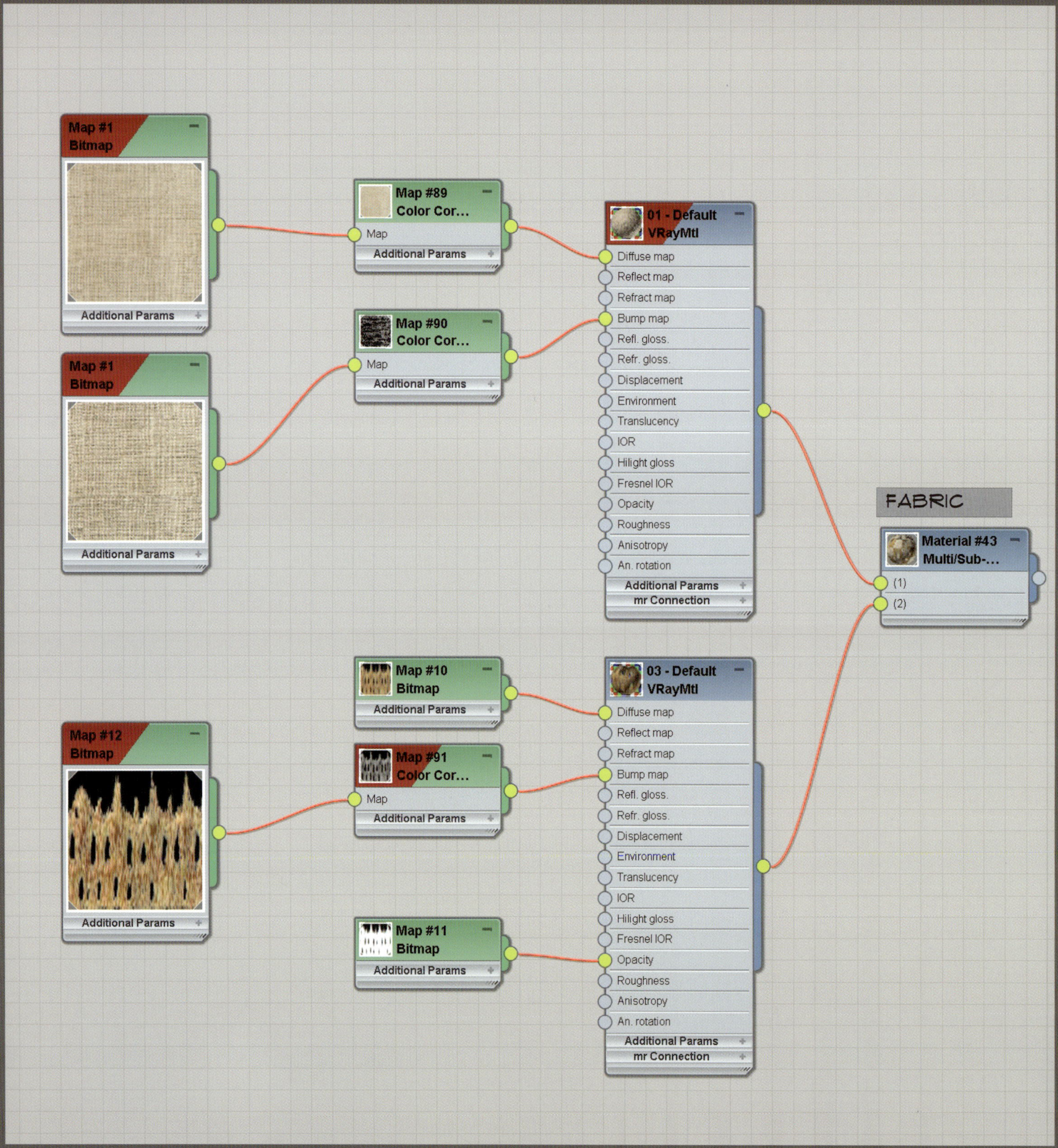

TABLE-WARE VOL I MADE WITH 3DS MAX AND V-RAY

SLATE MATERIAL EDITOR

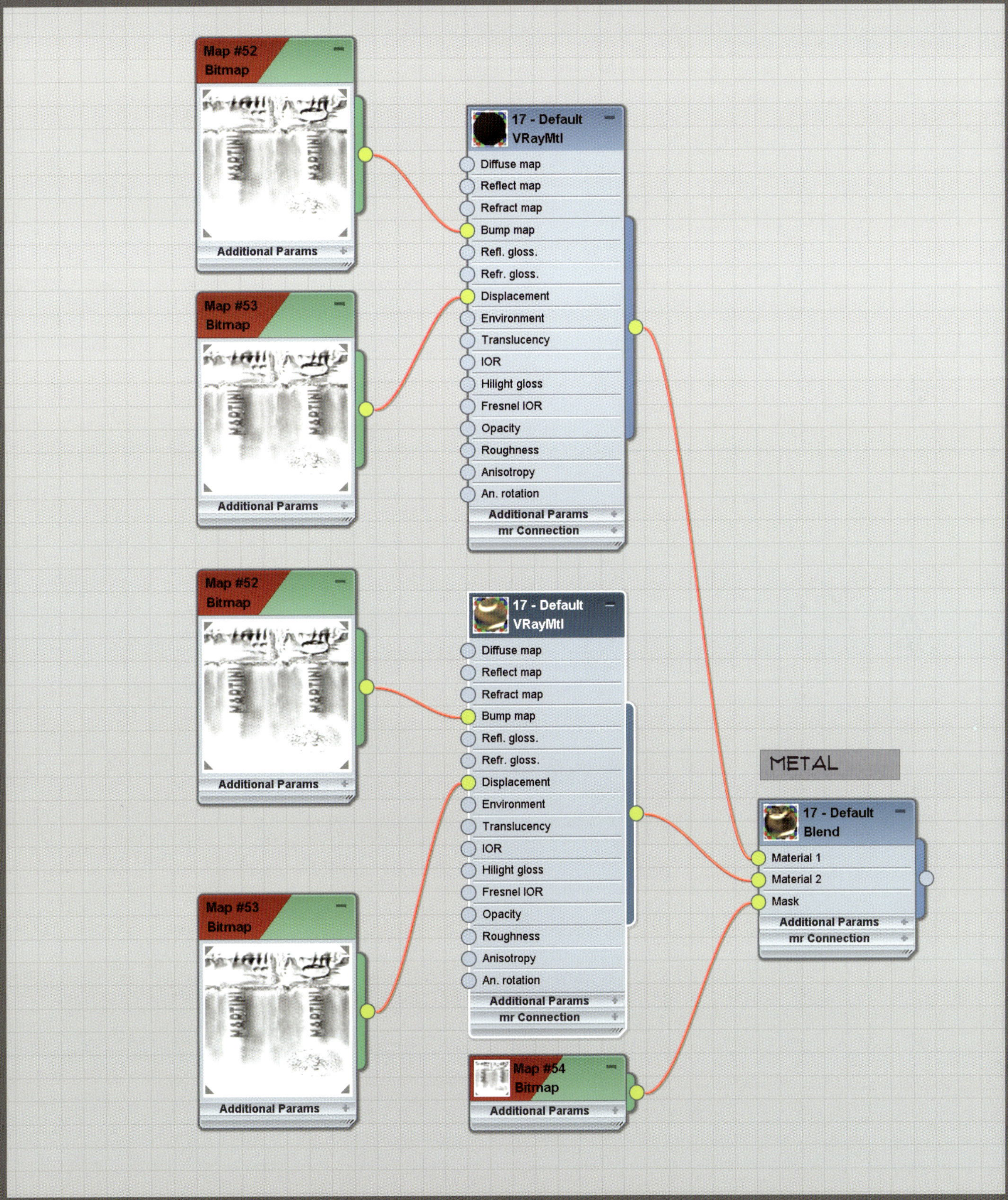

TABLE-WARE VOL I MADE WITH 3DS MAX AND V-RAY

TABLE-WARE 029

3DS MAX 2010 + FBX (V-RAY)   12,496 KB   CLASSICAL

MODIFIERS STACK
- NOT COLLAPSED
- COLLAPSED

TEXTURE
- NOT INCLUDED
- INCLUDED

MAPPING
- UNWRAP
- UVW MAP

TOTAL
- POLYS : 200,412
- VERTS : 130,124

TABLE_WARE

MAX

3DS

FBX – SIMPLE OBJECT WITHOUT MATERIALS (WITH MAPPING AND TEXTURES INCLUDED)

*.MAX – V-RAY 1.5 – OR HIGHER – WITH TEXTURES AND SHADERS

V-RAY – OBJECT PREPARED FOR V-RAY RENDERER (WITH TEXTURES AND SHADERS)

*. MAX 2010 – OR HIGHER

ELEMENT OF MODEL

PARTIAL GROUP

1. TABLE_WARE-A   2. TABLE_WARE-B   3. TABLE_WARE-C   4. TABLE_WARE-D

TABLE-WARE VOL 1 MADE WITH 3DS MAX AND V-RAY

SLATE MATERIAL EDITOR

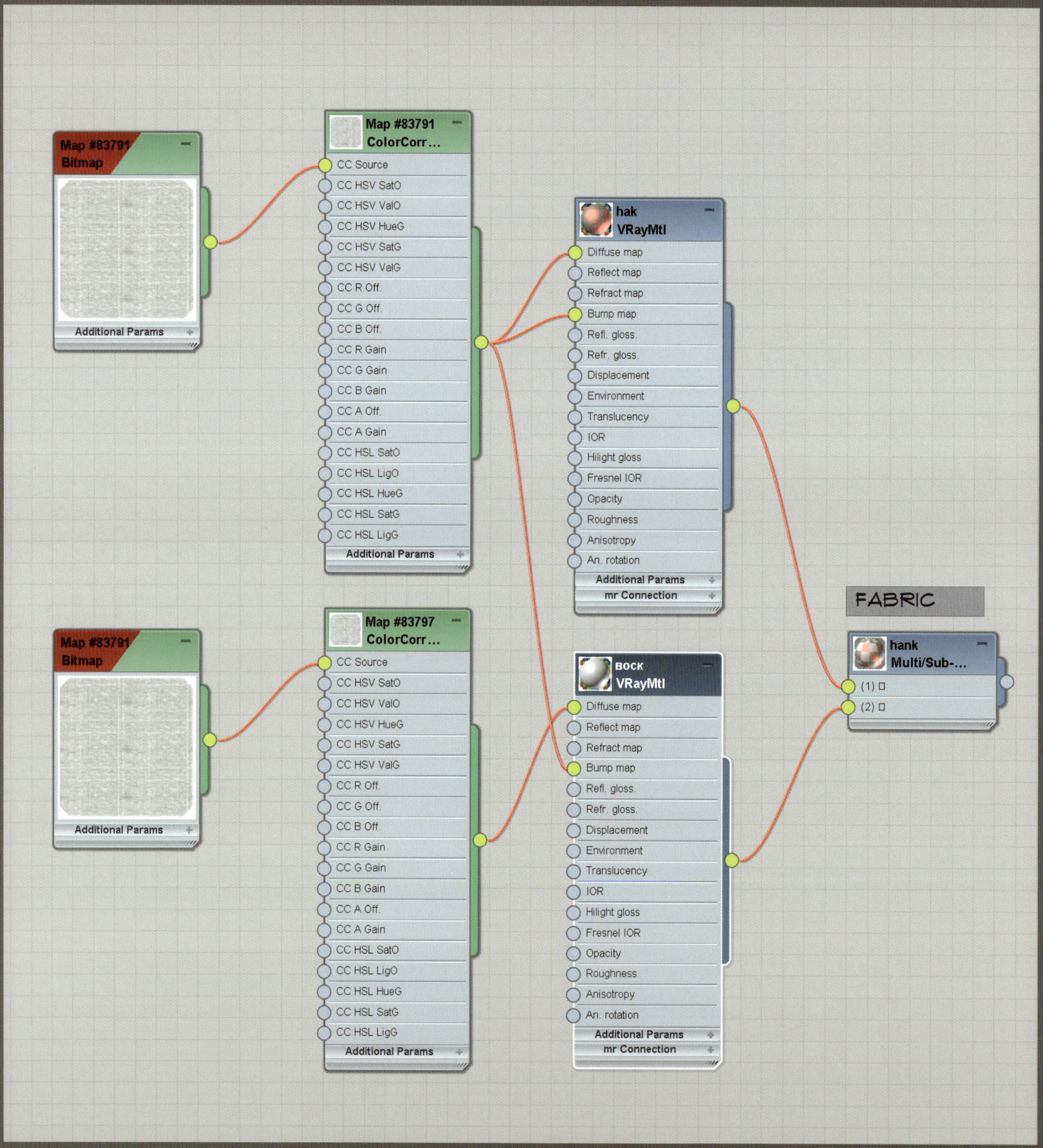

SLATE MATERIAL EDITOR

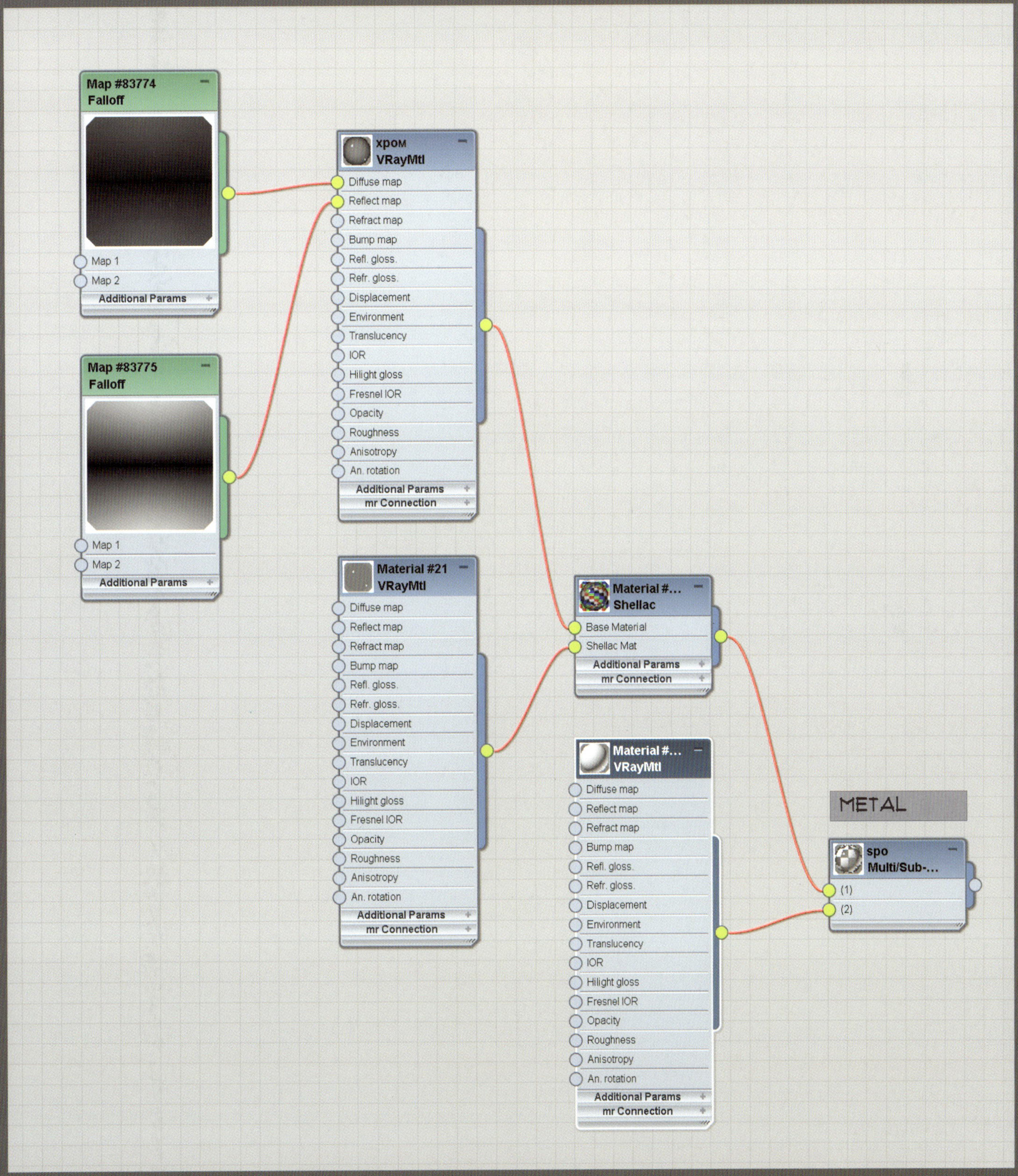

METAL

TABLE-WARE VOL I MADE WITH 3DS MAX AND V-RAY

| TABLE-WARE 030 | 3DS MAX 2010 + FBX (V-RAY) | 15,344 KB | CLASSICAL |

MODIFIERS STACK
NOT COLLAPSED
COLLAPSED

TEXTURE
NOT INCLUDED
INCLUDED

MAPPING
UNWRAP
UVW MAP

TOTAL
POLYS : 387,088
VERTS : 343,807

TABLE_WARE

MAX

3DS

FBX - SIMPLE OBJECT WITHOUT MATERIALS (WITH MAPPING AND TEXTURES INCLUDED)

*.MAX - V-RAY 1.5 - OR HIGHER - WITH TEXTURES AND SHADERS

V-RAY - OBJECT PREPARED FOR V-RAY RENDERER (WITH TEXTURES AND SHADERS)

*. MAX 2010 - OR HIGHER

ELEMENT OF MODEL

PARTIAL GROUP

1. TABLE_WARE-A   2. TABLE_WARE-B   3. TABLE_WARE-C   4. TABLE_WARE-D

TABLE-WARE VOL1 MADE WITH 3DS MAX AND V-RAY

SLATE MATERIAL EDITOR

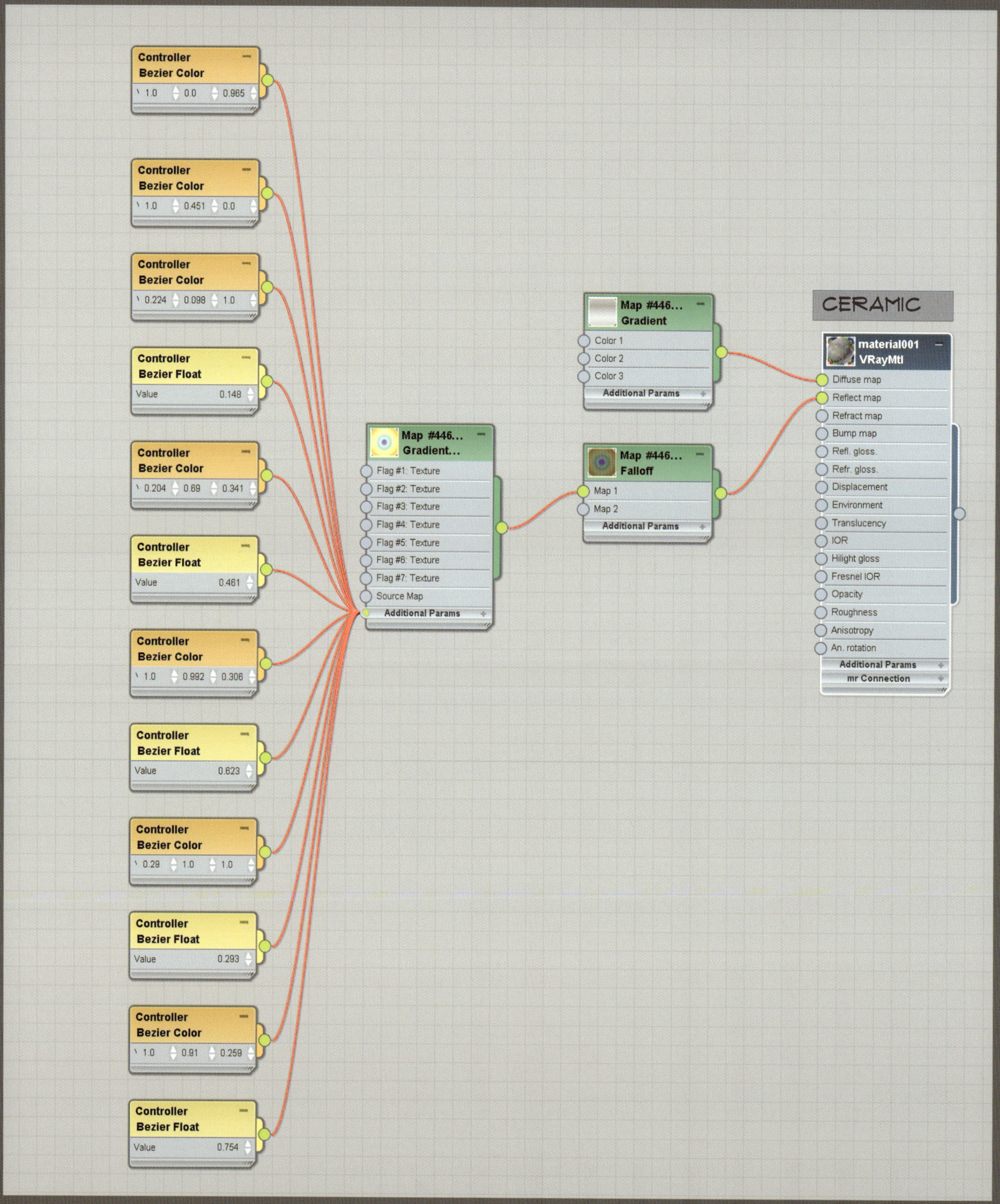

TABLE-WARE VOL I MADE WITH 3DS MAX AND V-RAY

SLATE MATERIAL EDITOR

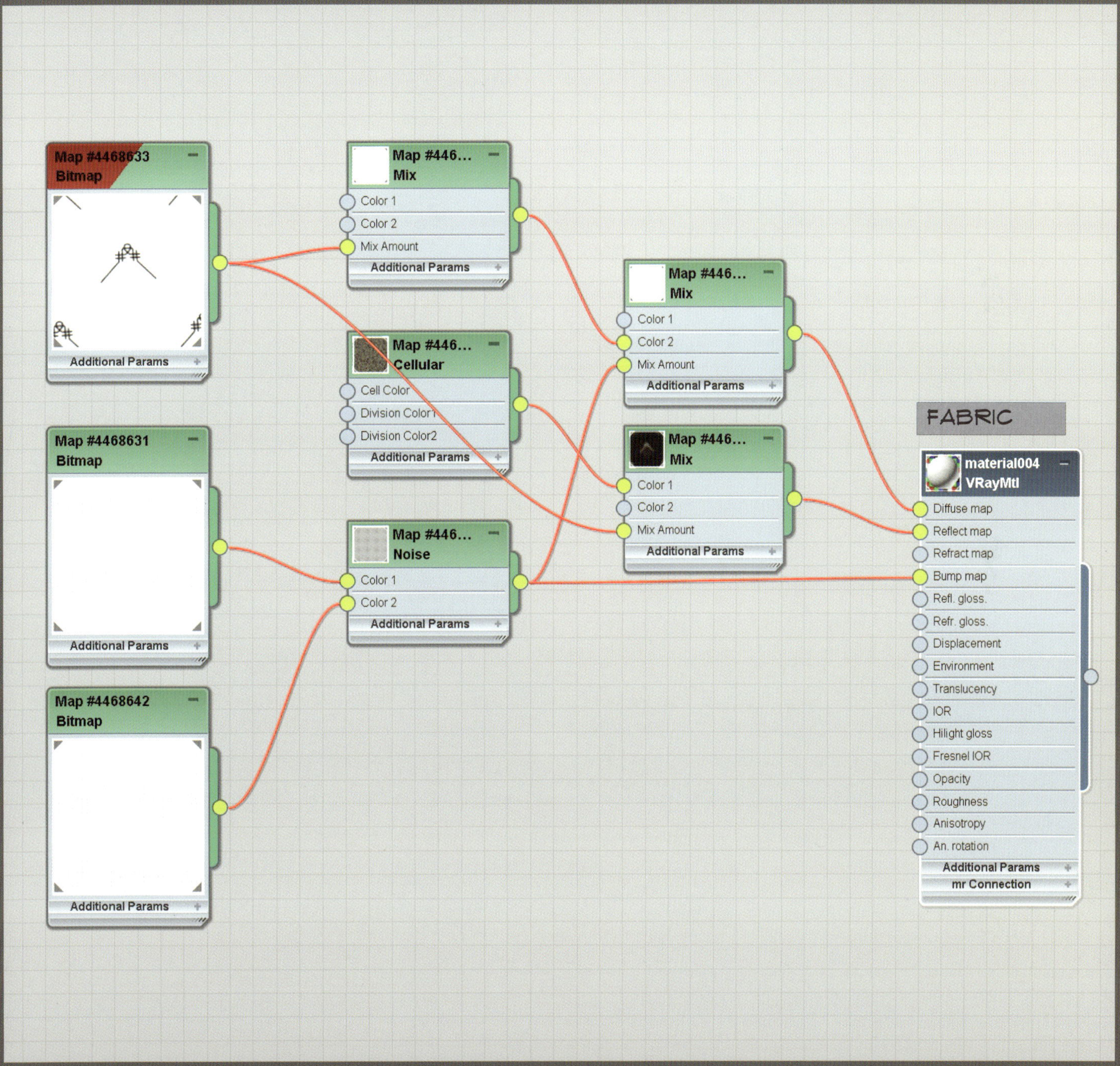

FABRIC

TABLE-WARE VOL 1 MADE WITH 3DS MAX AND V-RAY

| TABLE-WARE 031 | 3DS MAX 2010 + FBX (V-RAY) | 71,504 KB | CLASSICAL |

MODIFIERS STACK
- NOT COLLAPSED
- COLLAPSED

TEXTURE
- NOT INCLUDED
- INCLUDED

MAPPING
- UNWRAP
- UVW MAP

TOTAL
POLYS : 453,009
VERTS : 392,120

TABLE_WARE

MAX
3DS
FBX - SIMPLE OBJECT WITHOUT MATERIALS (WITH MAPPING AND TEXTURES INCLUDED)
*.MAX - V-RAY 1.5 - OR HIGHER - WITH TEXTURES AND SHADERS
V-RAY - OBJECT PREPARED FOR V-RAY RENDERER (WITH TEXTURES AND SHADERS)

*. MAX 2010 - OR HIGHER

MADE WITH:
THE NEW DIGITAL AGE

ELEMENT OF MODEL

PARTIAL GROUP

1. TABLE_WARE-A
2. TABLE_WARE-B
3. TABLE_WARE-C
4. TABLE_WARE-D

TABLE-WARE VOL I MADE WITH 3DS MAX AND V-RAY

SLATE MATERIAL EDITOR

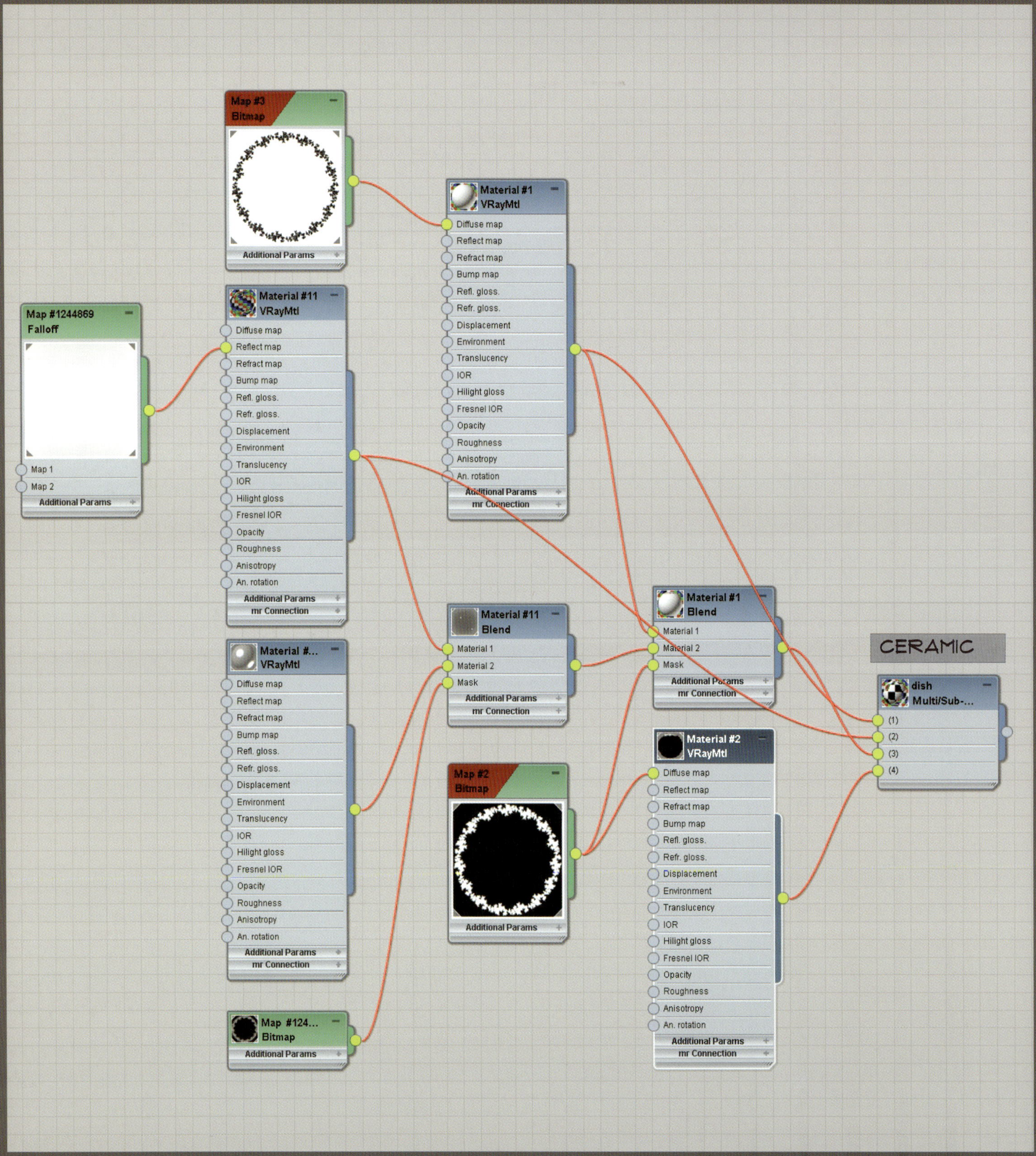

TABLE-WARE VOL I MADE WITH 3DS MAX AND V-RAY

SLATE MATERIAL EDITOR

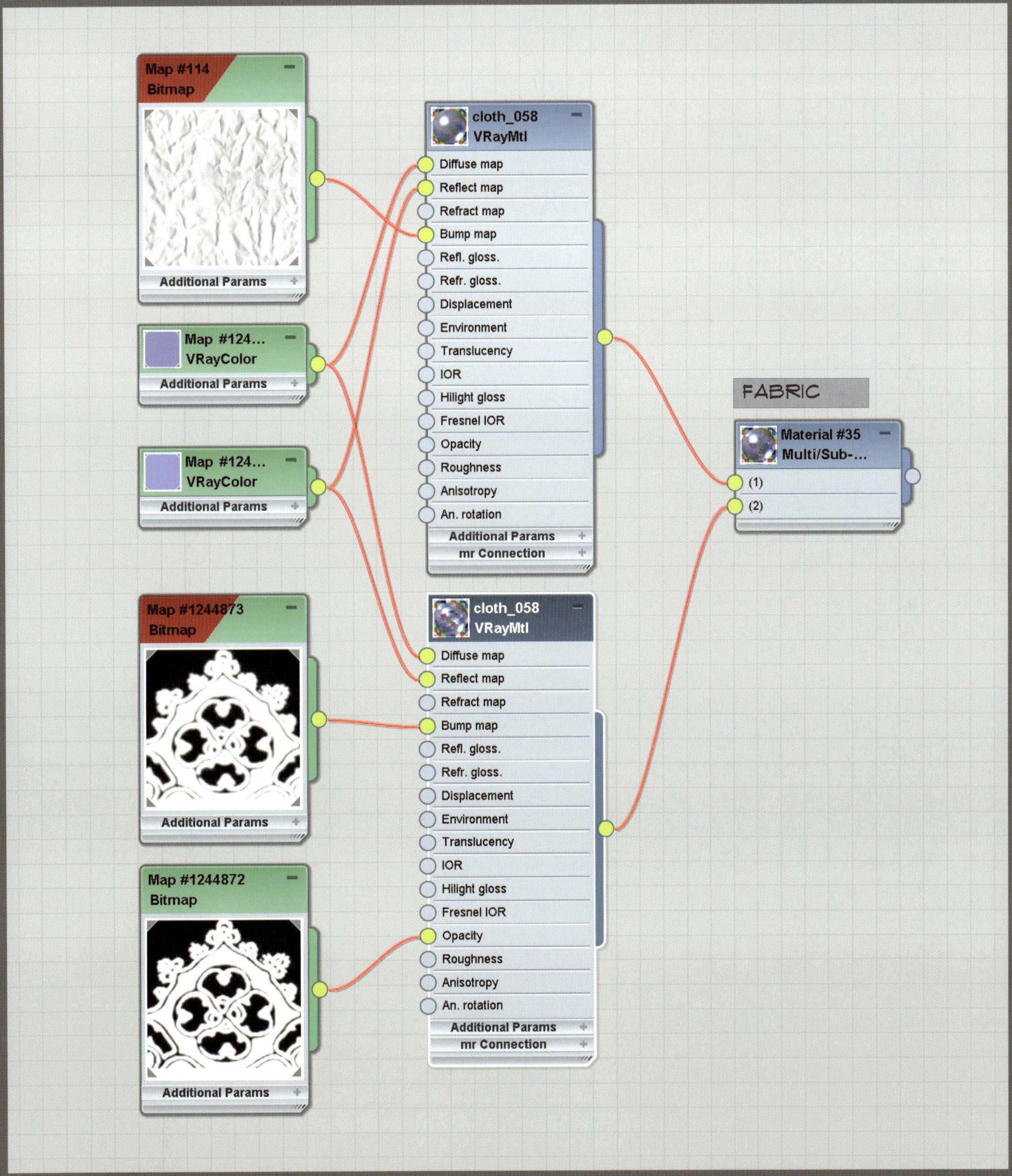

FABRIC

TABLE-WARE VOL I MADE WITH 3DS MAX AND V-RAY

| TABLE-WARE 032 | 3DS MAX 2010 + FBX (V-RAY) | 4,416 KB | CLASSICAL |

MODIFIERS STACK
NOT COLLAPSED
COLLAPSED

TEXTURE
NOT INCLUDED
INCLUDED

MAPPING
UNWRAP
UVW MAP

TOTAL
POLYS : 26,050
VERTS : 27,036

TABLE_WARE

MAX
3DS
FBX - SIMPLE OBJECT WITHOUT MATERIALS (WITH MAPPING AND TEXTURES INCLUDED)
*.MAX - V-RAY 1.5 - OR HIGHER - WITH TEXTURES AND SHADERS
V-RAY - OBJECT PREPARED FOR V-RAY RENDERER (WITH TEXTURES AND SHADERS)

*. MAX 2010 - OR HIGHER

ELEMENT OF MODEL

PARTIAL GROUP

1. TABLE_WARE-A   2. TABLE_WARE-B   3. TABLE_WARE-C   4. TABLE_WARE-D

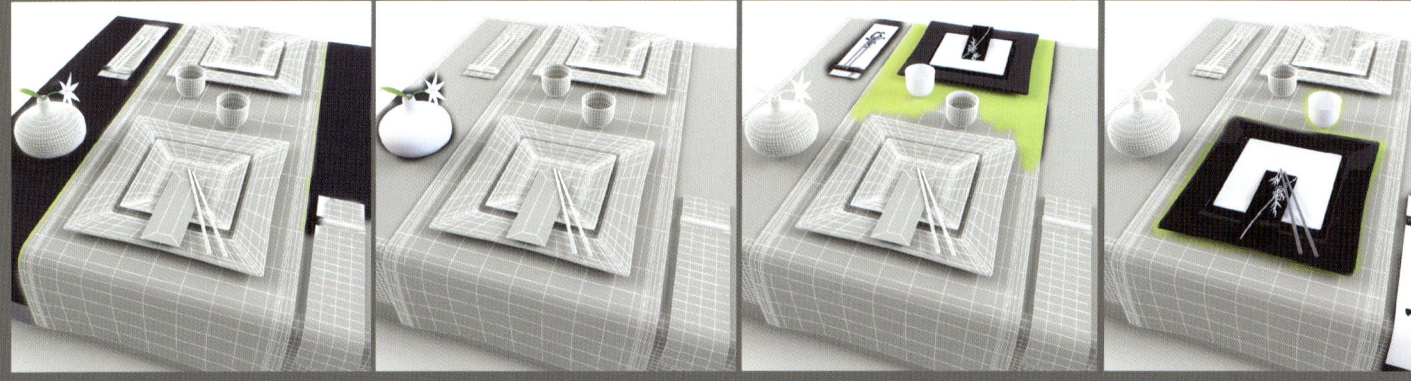

TABLE-WARE VOL I MADE WITH 3DS MAX AND V-RAY

SLATE MATERIAL EDITOR

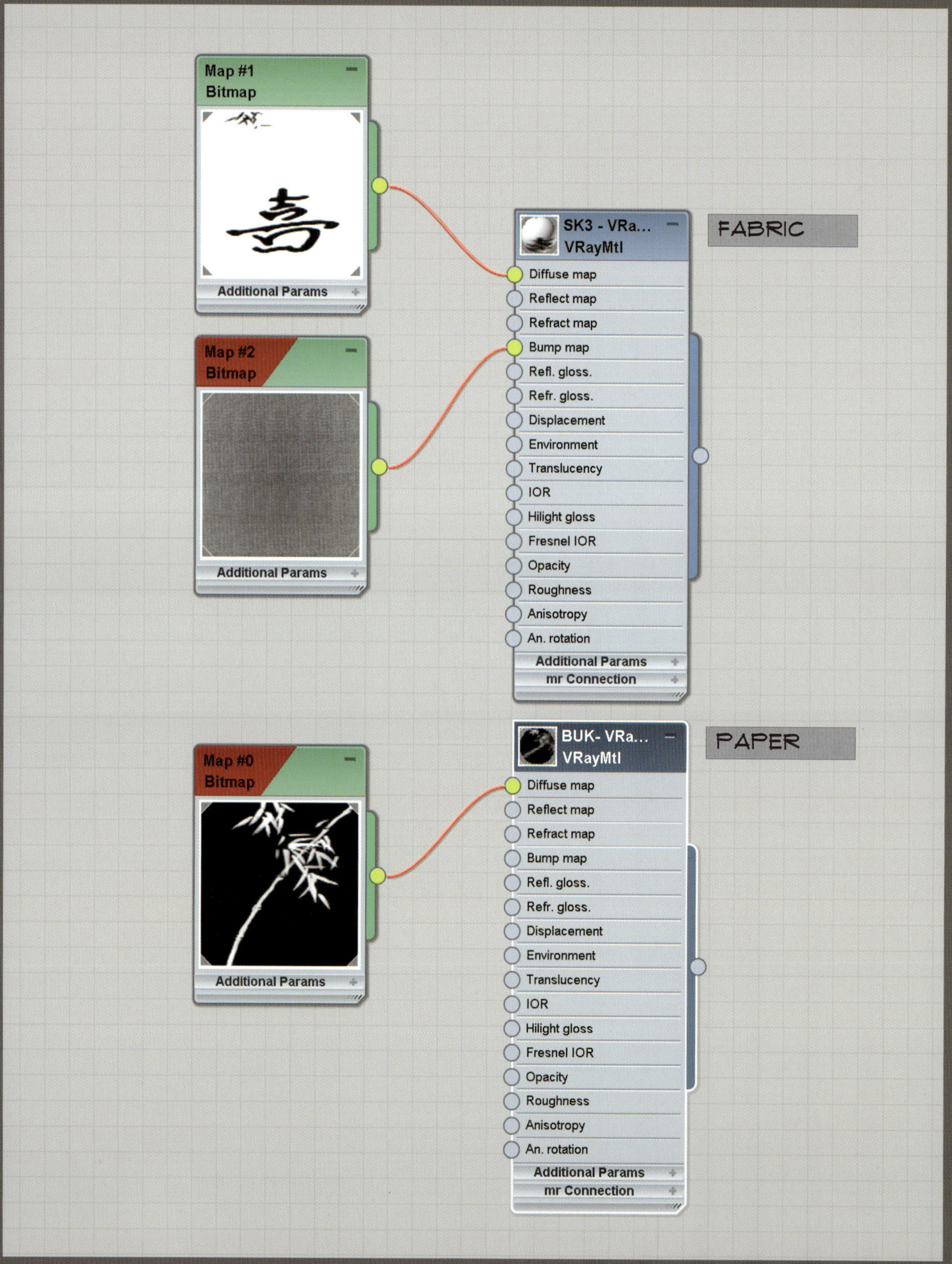

SLATE MATERIAL EDITOR

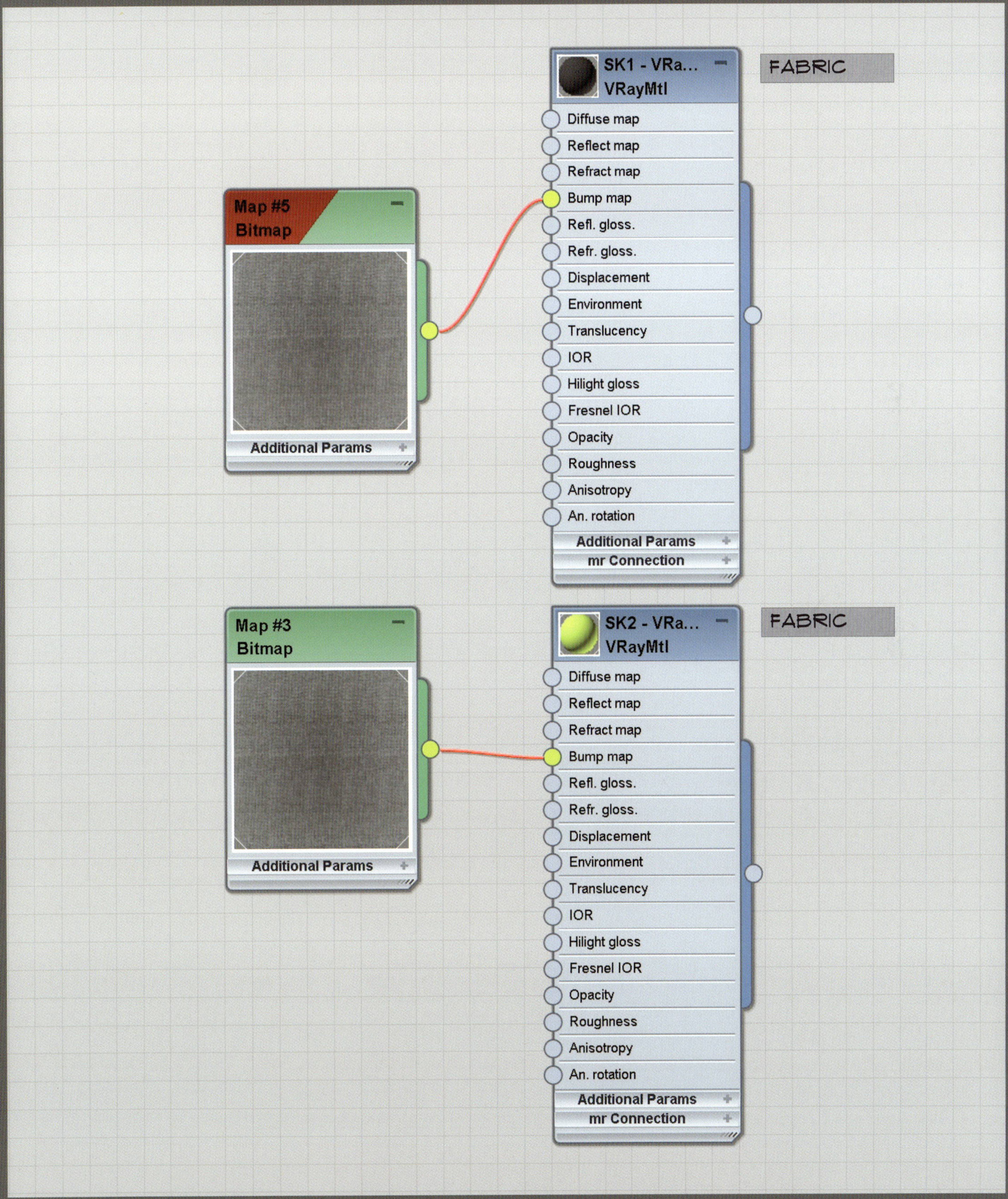

| TABLE-WARE 033 | 3DS MAX 2010 + FBX (V-RAY) |  62,804 KB | ☆ MODERN |

MODIFIERS STACK
NOT COLLAPSED
COLLAPSED

TEXTURE
NOT INCLUDED
INCLUDED

MAPPING
UNWRAP
UVW MAP

TOTAL
POLYS : 439,307
VERTS : 400,194

TABLE_WARE

MAX

3DS

FBX - SIMPLE OBJECT WITHOUT MATERIALS (WITH MAPPING AND TEXTURES INCLUDED)

*.MAX - V-RAY 1.5 - OR HIGHER - WITH TEXTURES AND SHADERS

V-RAY - OBJECT PREPARED FOR V-RAY RENDERER (WITH TEXTURES AND SHADERS)

*. MAX 2010 - OR HIGHER

ELEMENT OF MODEL

PARTIAL GROUP

1. TABLE_WARE-A   2. TABLE_WARE-B   3. TABLE_WARE-C   4. TABLE_WARE-D

TABLE-WARE VOL.1 MADE WITH 3DS MAX AND V-RAY

SLATE MATERIAL EDITOR

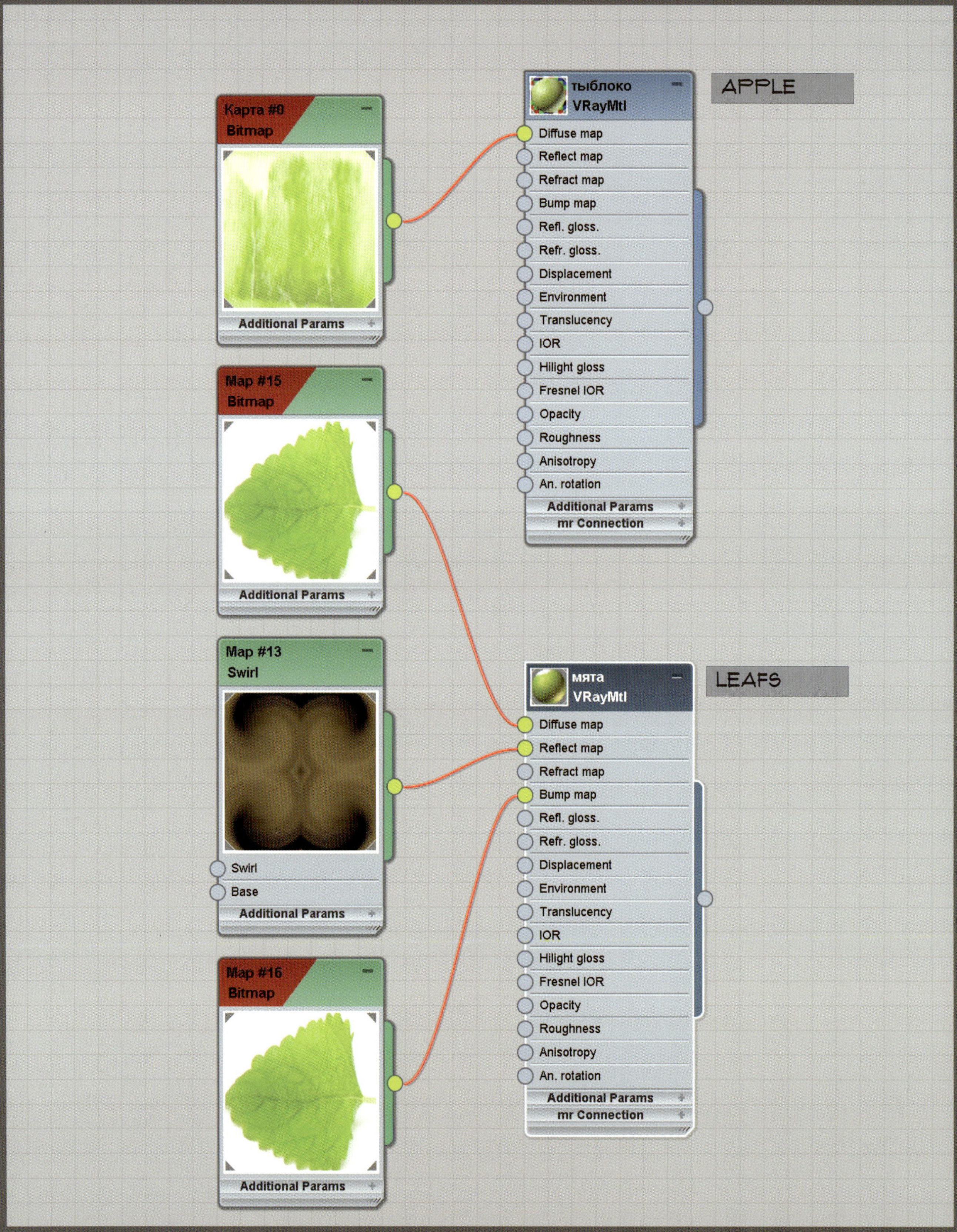

TABLE-WARE VOL I MADE WITH 3DS MAX AND V-RAY

SLATE MATERIAL EDITOR

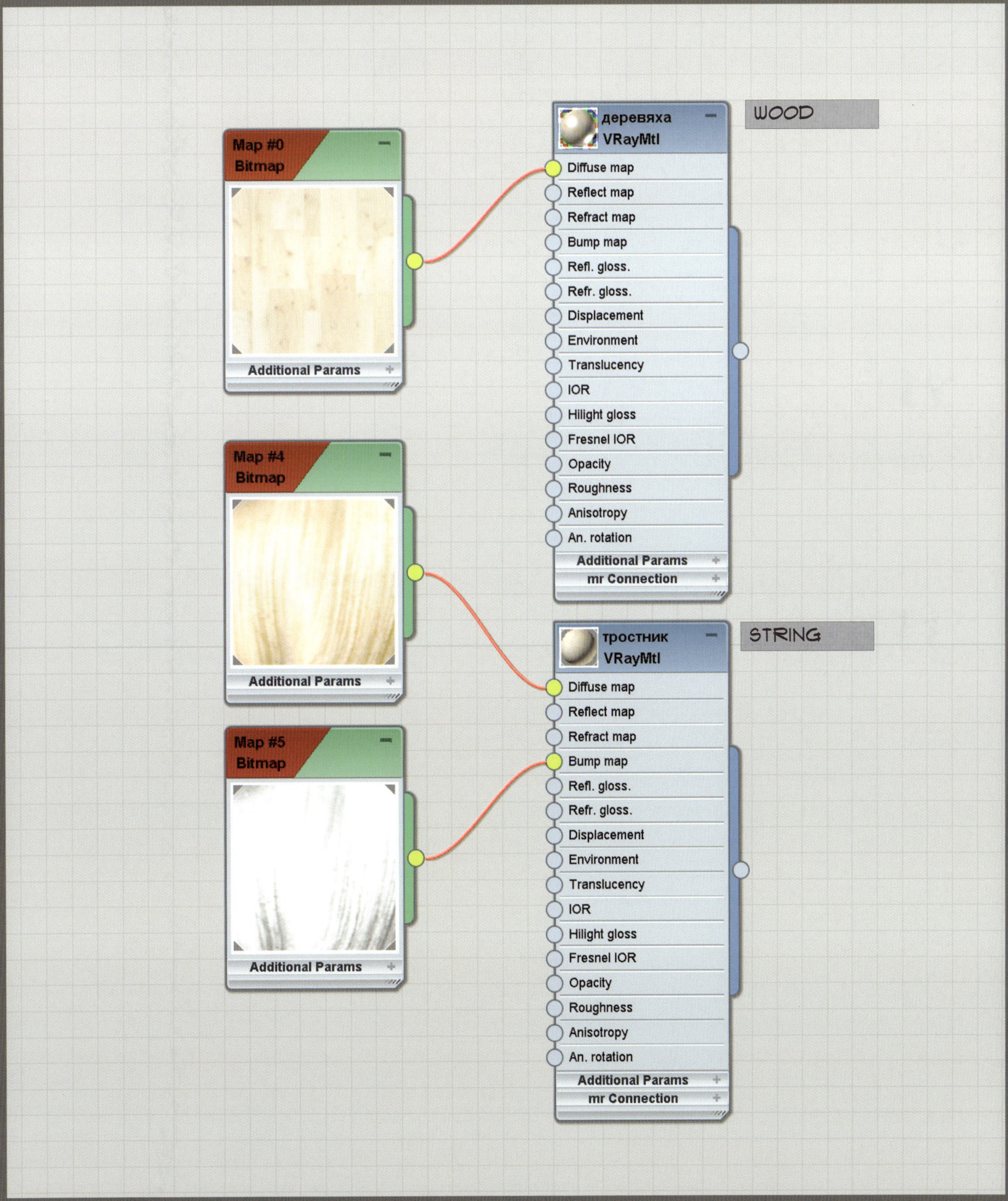

TABLE-WARE VOL I MADE WITH 3DS MAX AND V-RAY

| TABLE-WARE 034 | 3DS MAX 2010 + FBX (V-RAY) | 27,936 KB | CLASSICAL |

MODIFIERS STACK
NOT COLLAPSED
COLLAPSED

TEXTURE
NOT INCLUDED
INCLUDED

MAPPING
UNWRAP
UVW MAP

TOTAL
POLYS : 198,531
VERTS : 242,653

TABLE_WARE

MAX
3DS
FBX – SIMPLE OBJECT WITHOUT MATERIALS (WITH MAPPING AND TEXTURES INCLUDED)
*.MAX – V-RAY 1.5 – OR HIGHER – WITH TEXTURES AND SHADERS
V-RAY – OBJECT PREPARED FOR V-RAY RENDERER (WITH TEXTURES AND SHADERS)

*. MAX 2010 – OR HIGHER

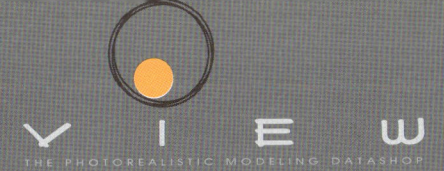

ELEMENT OF MODEL

PARTIAL GROUP

1. TABLE_WARE-A   2. TABLE_WARE-B   3. TABLE_WARE-C   4. TABLE_WARE-D

TABLE-WARE VOL 1 MADE WITH 3DS MAX AND V-RAY

SLATE MATERIAL EDITOR

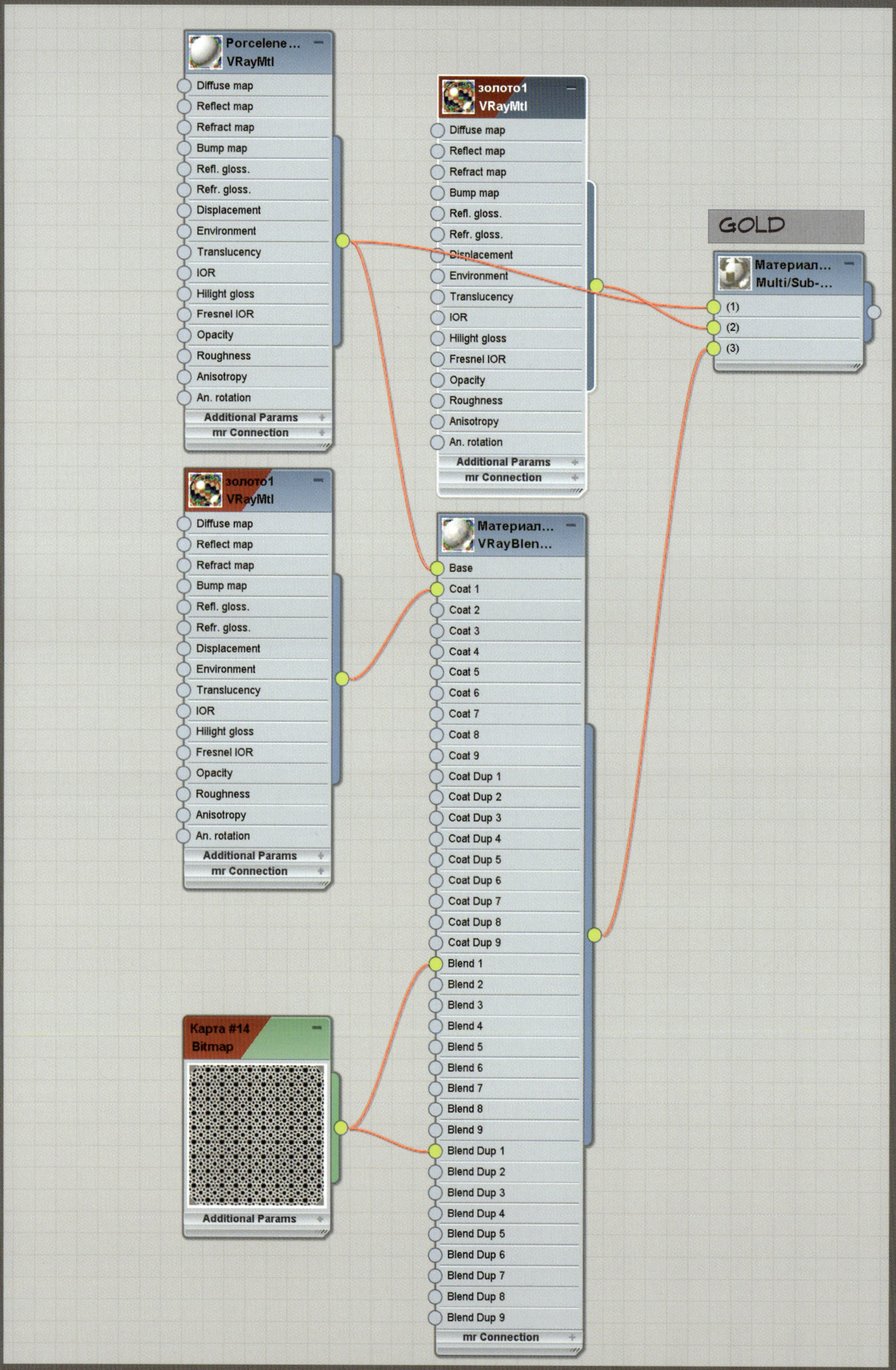

TABLE-WARE VOL I MADE WITH 3DS MAX AND V-RAY

SLATE MATERIAL EDITOR

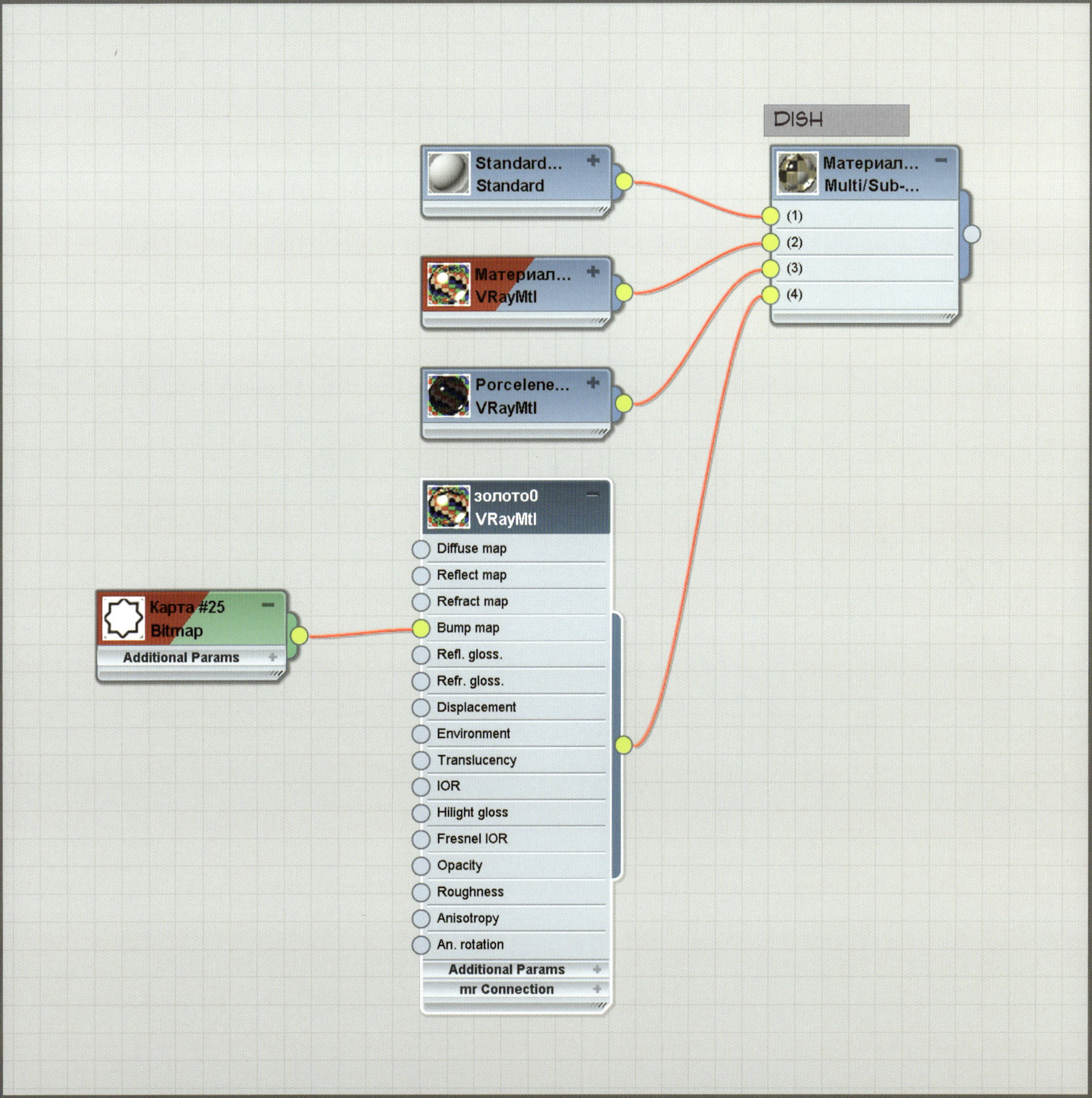

TABLE-WARE VOL I MADE WITH 3DS MAX AND V-RAY

TABLE-WARE 035

3DS MAX 2010 + FBX (V-RAY)   26,428 KB   MODERN

MODIFIERS STACK
- NOT COLLAPSED
- COLLAPSED

TEXTURE
- NOT INCLUDED
- INCLUDED

MAPPING
- UNWRAP
- UVW MAP

TOTAL
- POLYS : 577,155
- VERTS : 361,857

TABLE_WARE

MAX

3DS

FBX – SIMPLE OBJECT WITHOUT MATERIALS (WITH MAPPING AND TEXTURES INCLUDED)

*.MAX – V-RAY 1.5 – OR HIGHER – WITH TEXTURES AND SHADERS

V-RAY – OBJECT PREPARED FOR V-RAY RENDERER (WITH TEXTURES AND SHADERS)

*. MAX 2010 – OR HIGHER

ELEMENT OF MODEL

PARTIAL GROUP

1. TABLE_WARE-A
2. TABLE_WARE-B
3. TABLE_WARE-C
4. TABLE_WARE-D

TABLE-WARE VOL I MADE WITH 3DS MAX AND V-RAY

SLATE MATERIAL EDITOR

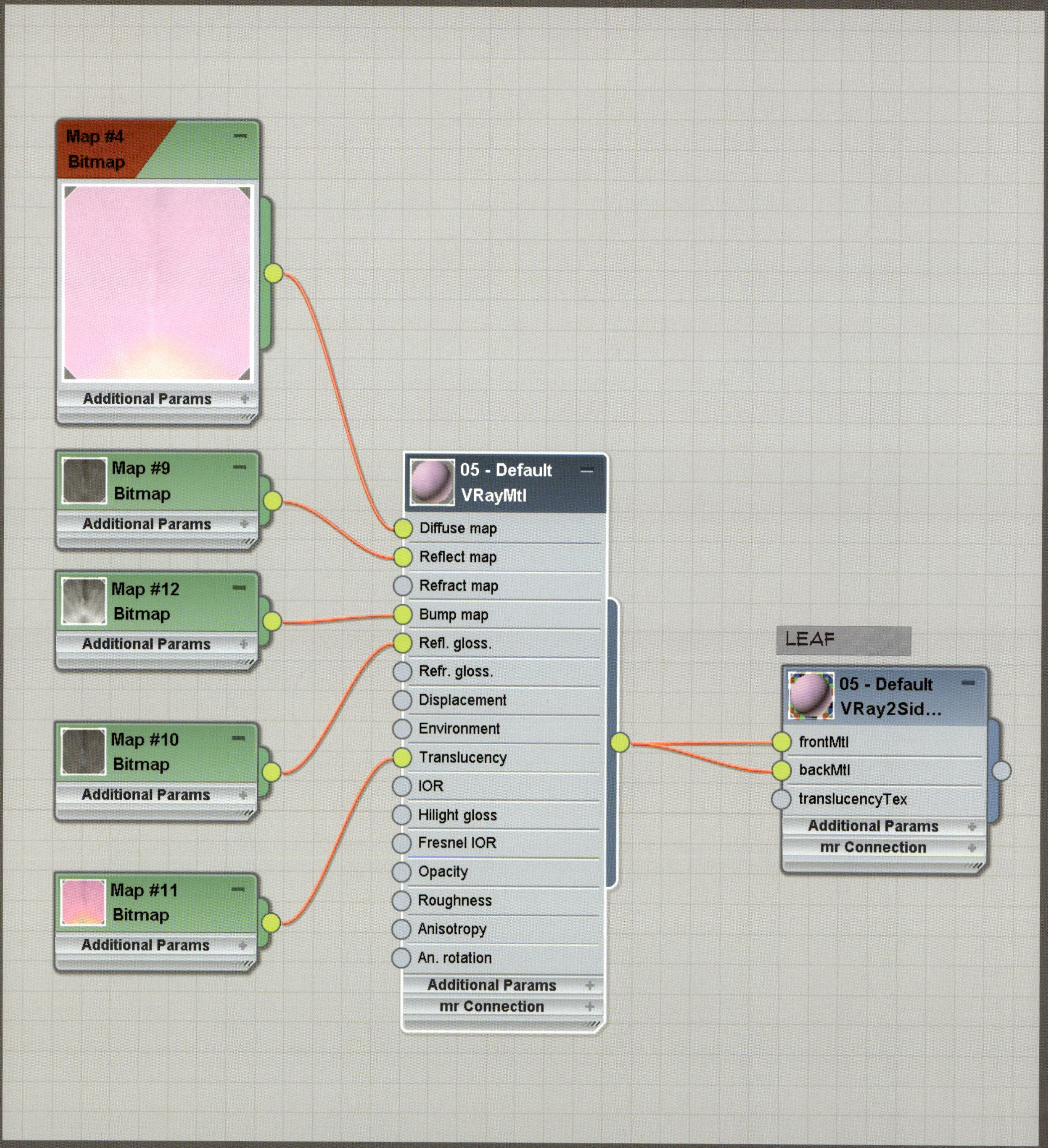

SLATE MATERIAL EDITOR

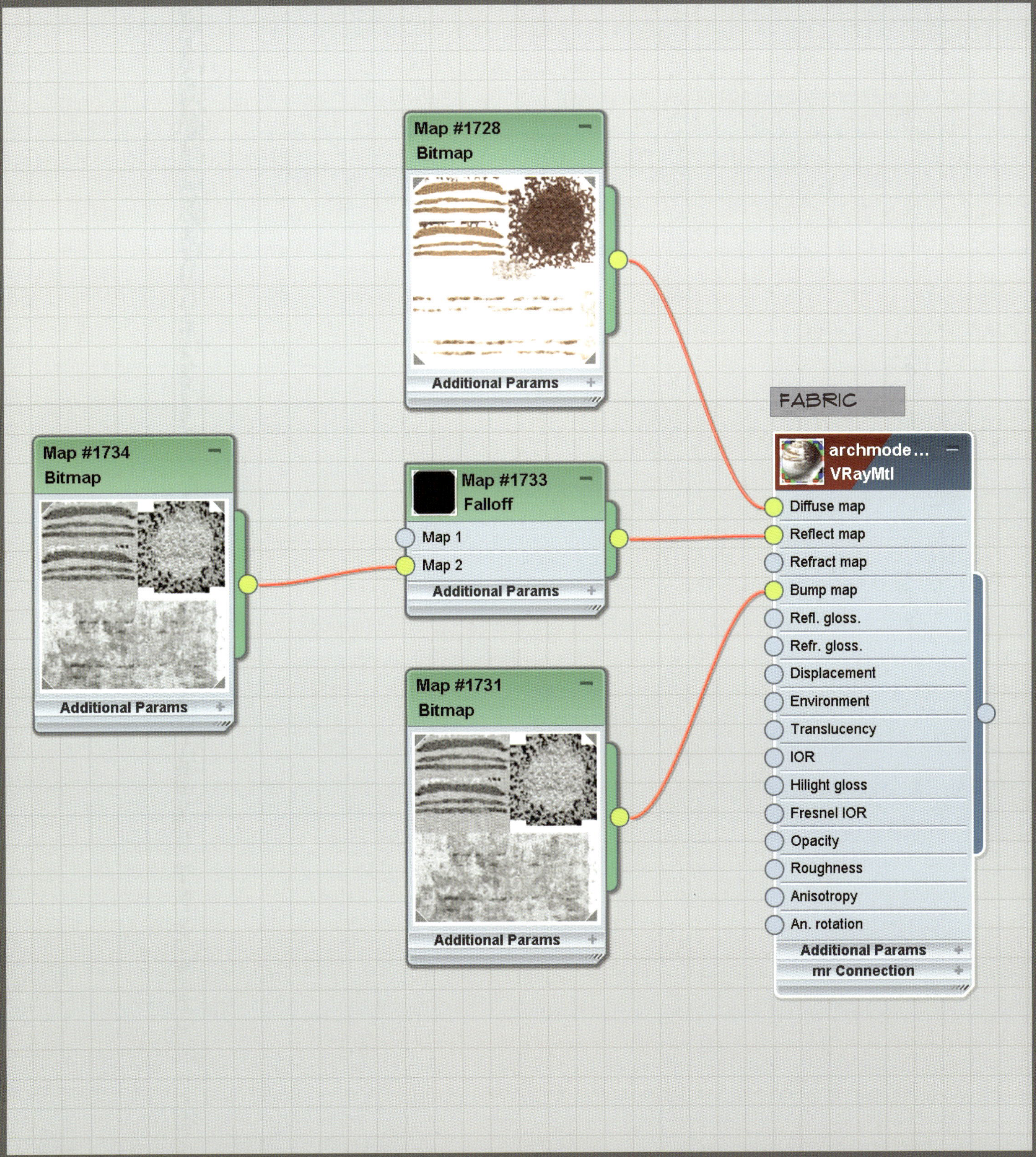

TABLE-WARE VOL I MADE WITH 3DS MAX AND V-RAY

TABLE-WARE 036

3DS MAX 2010 + FBX (V-RAY)   20,020 KB   CLASSICAL

MODIFIERS STACK
- NOT COLLAPSED
- COLLAPSED

TEXTURE
- NOT INCLUDED
- INCLUDED

MAPPING
- UNWRAP
- UVW MAP

TOTAL
- POLYS : 142,621
- VERTS : 152,013

TABLE_WARE

MAX

3DS

FBX - SIMPLE OBJECT WITHOUT MATERIALS (WITH MAPPING AND TEXTURES INCLUDED)

*.MAX - V-RAY 1.5 - OR HIGHER - WITH TEXTURES AND SHADERS

V-RAY - OBJECT PREPARED FOR V-RAY RENDERER (WITH TEXTURES AND SHADERS)

*. MAX 2010 - OR HIGHER

MADE WITH:
THE NEW DIGITAL AGE

ELEMENT OF MODEL

PARTIAL GROUP

1. TABLE_WARE-A   2. TABLE_WARE-B   3. TABLE_WARE-C   4. TABLE_WARE-D

TABLE-WARE VOL I MADE WITH 3DS MAX AND V-RAY

SLATE MATERIAL EDITOR

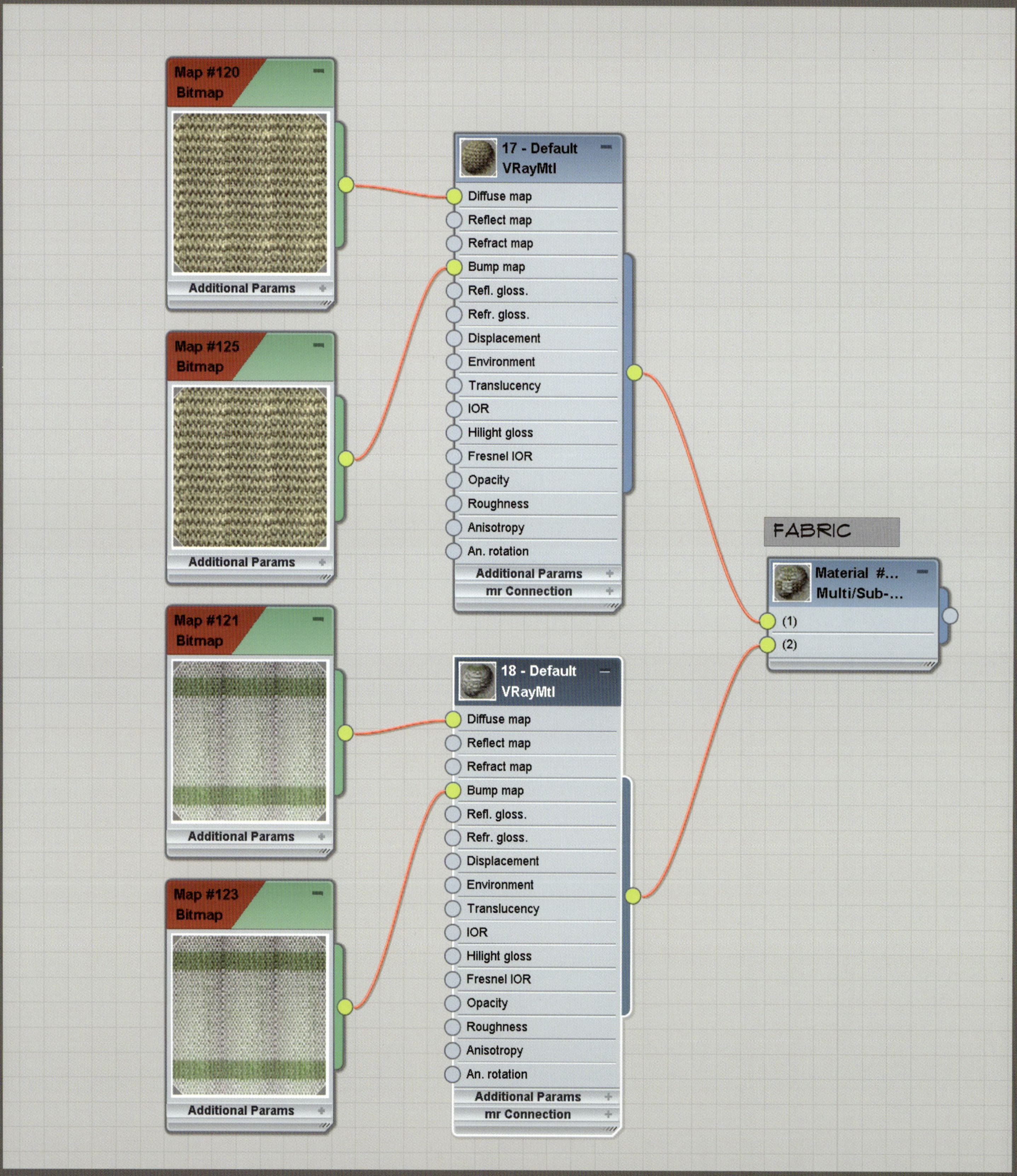

MADE WITH: 3ds Max + V-Ray
THE NEW DIGITAL AGE

SLATE MATERIAL EDITOR

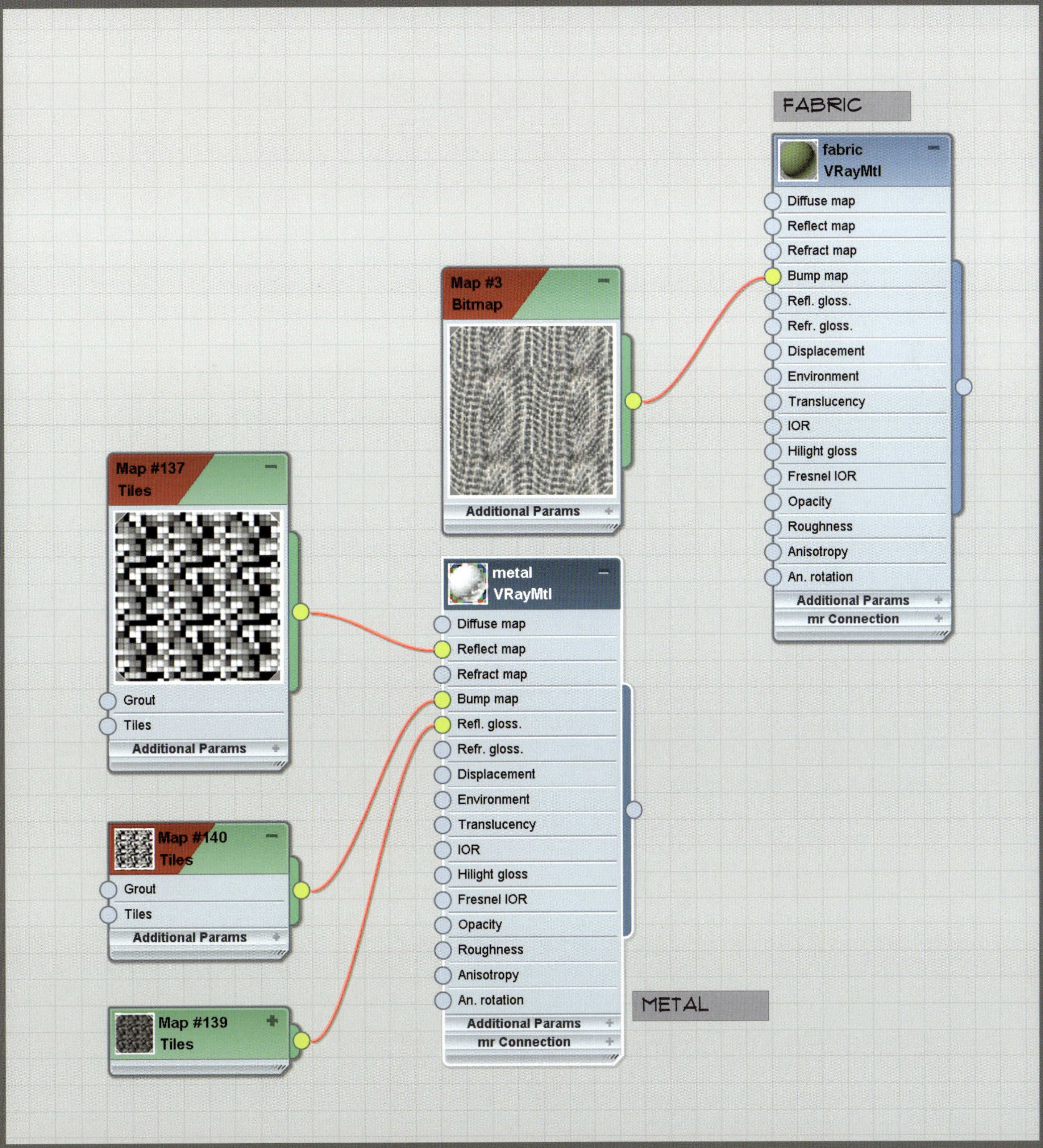

TABLE-WARE VOL I MADE WITH 3DS MAX AND V-RAY

| TABLE-WARE 037 | 3DS MAX 2010 + FBX (V-RAY) | 65,016 KB | CLASSICAL |

MODIFIERS STACK
NOT COLLAPSED
COLLAPSED

TEXTURE
NOT INCLUDED
INCLUDED

MAPPING
UNWRAP
UVW MAP

TOTAL
POLYS : 455,260
VERTS : 420,249

TABLE_WARE

MAX
3DS
FBX — SIMPLE OBJECT WITHOUT MATERIALS (WITH MAPPING AND TEXTURES INCLUDED)
*.MAX — V-RAY 1.5 — OR HIGHER — WITH TEXTURES AND SHADERS
V-RAY — OBJECT PREPARED FOR V-RAY RENDERER (WITH TEXTURES AND SHADERS)

*. MAX 2010 — OR HIGHER

ELEMENT OF MODEL

PARTIAL GROUP

1. TABLE_WARE-A   2. TABLE_WARE-B   3. TABLE_WARE-C   4. TABLE_WARE-D

TABLE-WARE VOL 1 MADE WITH 3DS MAX AND V-RAY

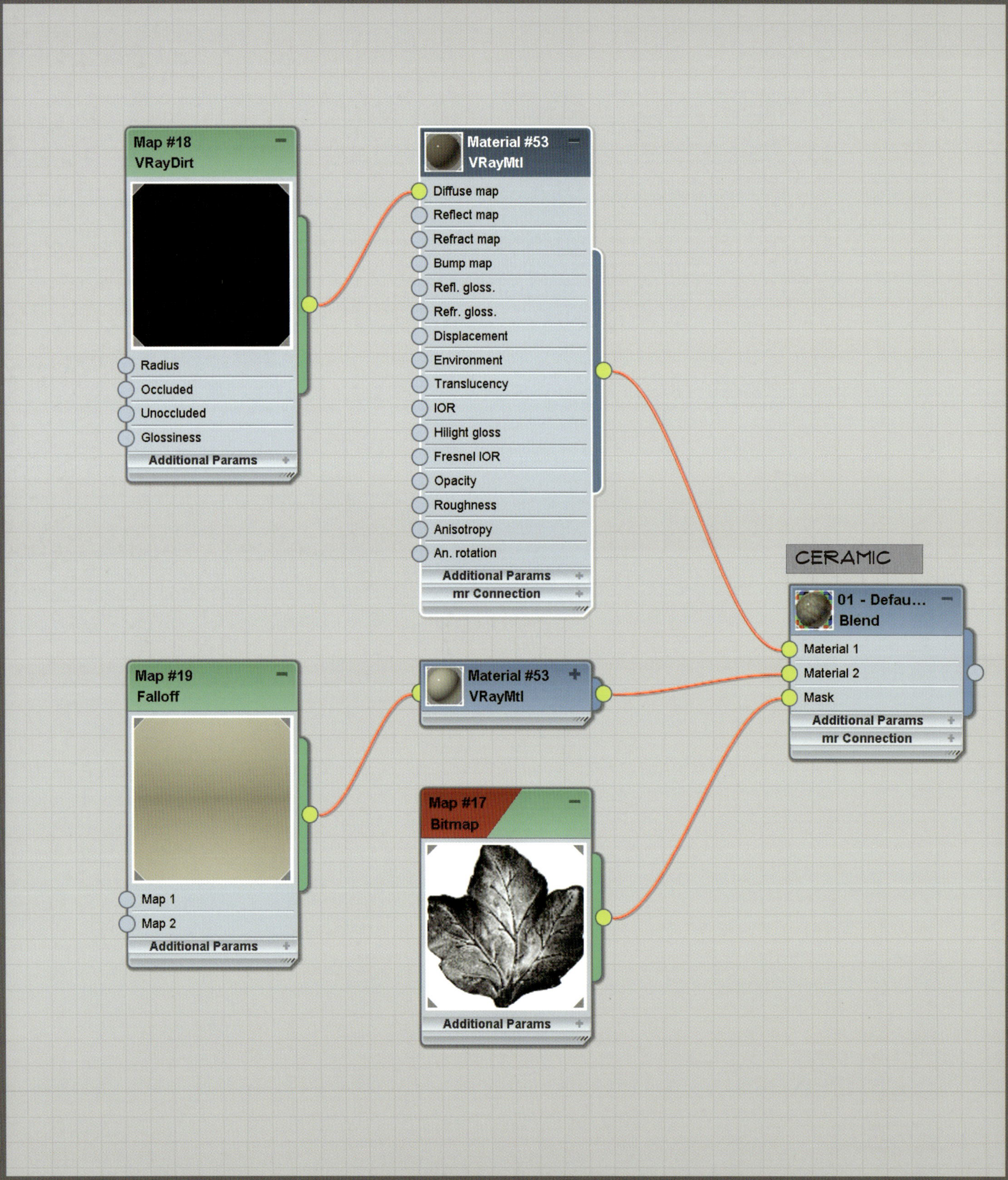

SLATE MATERIAL EDITOR

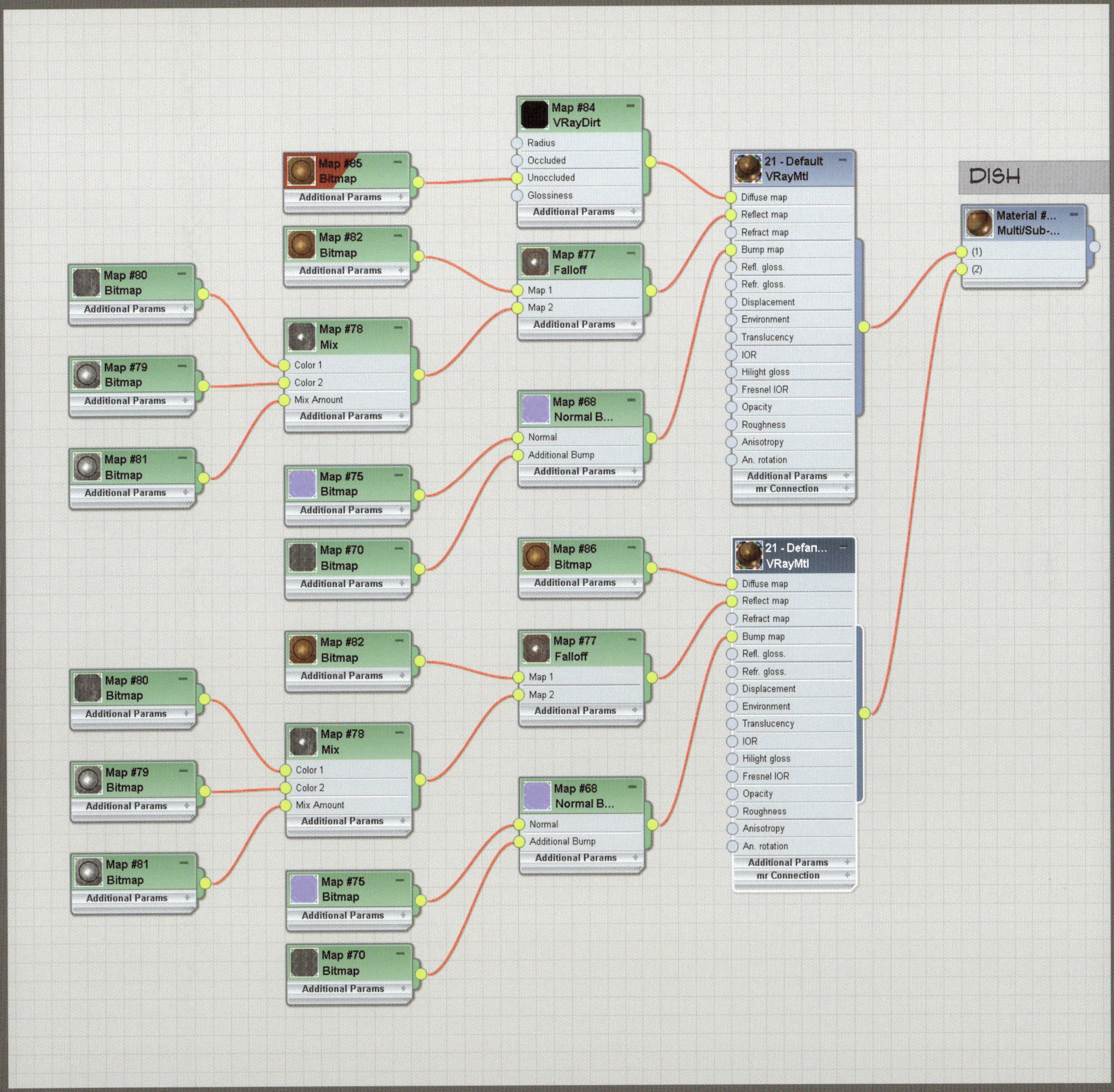

TABLE-WARE VOL I MADE WITH 3DS MAX AND V-RAY

| TABLE-WARE 038 | 3DS MAX 2010 + FBX (V-RAY) | 47,088 KB | CLASSICAL |

MODIFIERS STACK
NOT COLLAPSED
COLLAPSED

TEXTURE
NOT INCLUDED
INCLUDED

MAPPING
UNWRAP
UVW MAP

TOTAL
POLYS : 408,511
VERTS : 333,566

TABLE_WARE

MAX

3DS

FBX – SIMPLE OBJECT WITHOUT MATERIALS (WITH MAPPING AND TEXTURES INCLUDED)

*.MAX – V-RAY 1.5 – OR HIGHER – WITH TEXTURES AND SHADERS

V-RAY – OBJECT PREPARED FOR V-RAY RENDERER (WITH TEXTURES AND SHADERS)

*. MAX 2010 – OR HIGHER

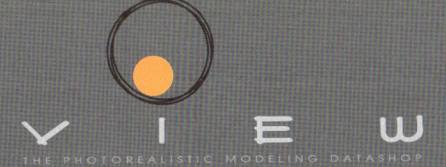

ELEMENT OF MODEL

PARTIAL GROUP

1. TABLE_WARE-A
2. TABLE_WARE-B
3. TABLE_WARE-C
4. TABLE_WARE-D

TABLE-WARE VOL 1 MADE WITH 3DS MAX AND V-RAY

SLATE MATERIAL EDITOR

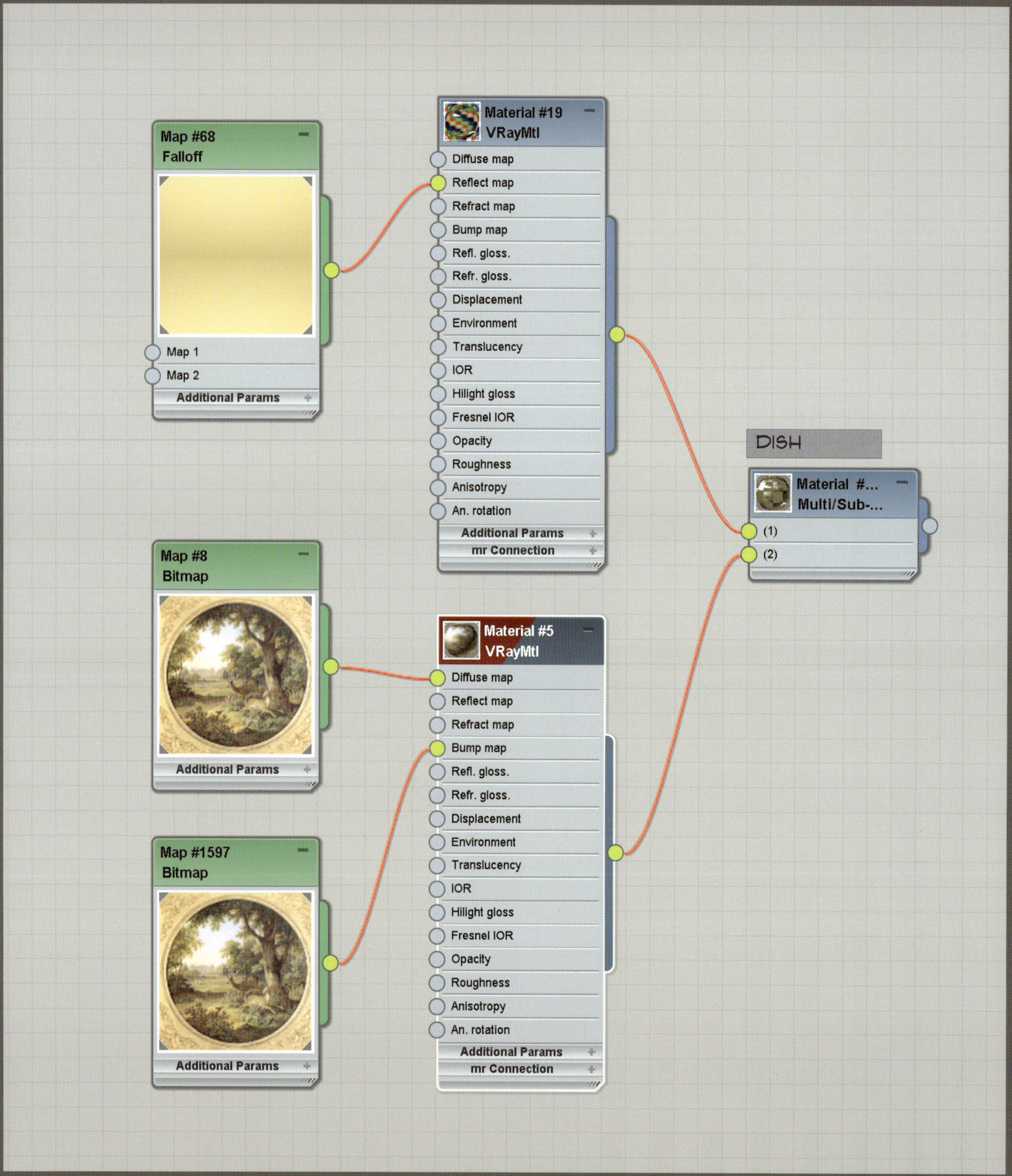

211 / 212

TABLE-WARE VOL I MADE WITH 3DS MAX AND V-RAY

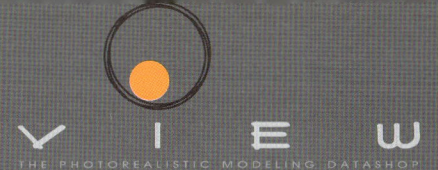

SLATE MATERIAL EDITOR

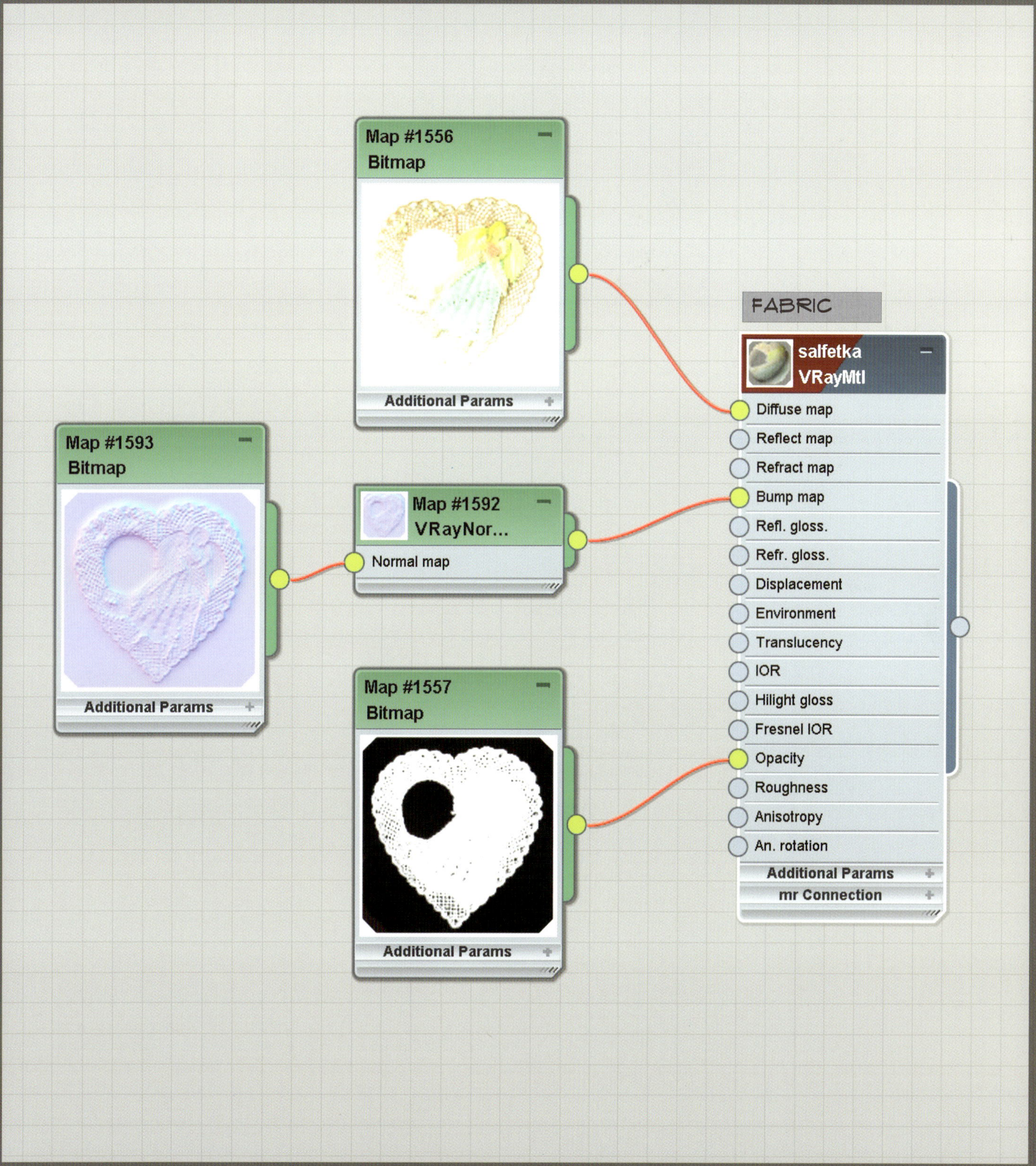

TABLE-WARE VOL 1 MADE WITH 3DS MAX AND V-RAY

| TABLE-WARE 039 | 3DS MAX 2010 + FBX (V-RAY) | 17,188 KB | MODERN |

MODIFIERS STACK
- NOT COLLAPSED
- COLLAPSED

TEXTURE
- NOT INCLUDED
- INCLUDED

MAPPING
- UNWRAP
- UVW MAP

TOTAL
- POLYS : 171,104
- VERTS : 120,884

TABLE_WARE

MAX
3DS
FBX - SIMPLE OBJECT WITHOUT MATERIALS (WITH MAPPING AND TEXTURES INCLUDED)
*.MAX - V-RAY 1.5 - OR HIGHER - WITH TEXTURES AND SHADERS
V-RAY - OBJECT PREPARED FOR V-RAY RENDERER (WITH TEXTURES AND SHADERS)

*. MAX 2010 - OR HIGHER

ELEMENT OF MODEL

PARTIAL GROUP

1. TABLE_WARE-A   2. TABLE_WARE-B   3. TABLE_WARE-C   4. TABLE_WARE-D

TABLE-WARE VOL 1 MADE WITH 3DS MAX AND V-RAY

SLATE MATERIAL EDITOR

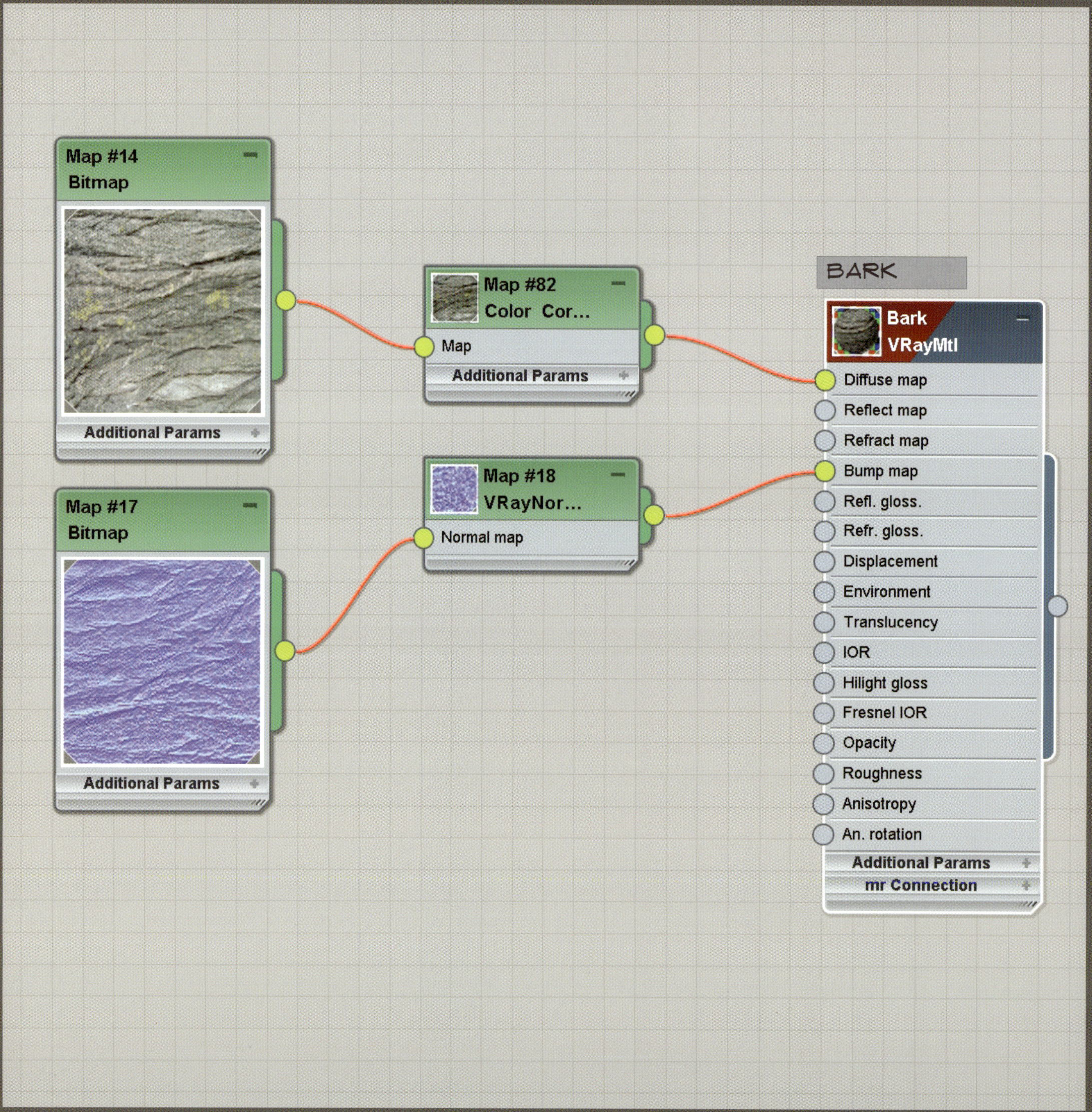

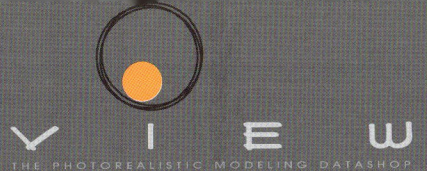

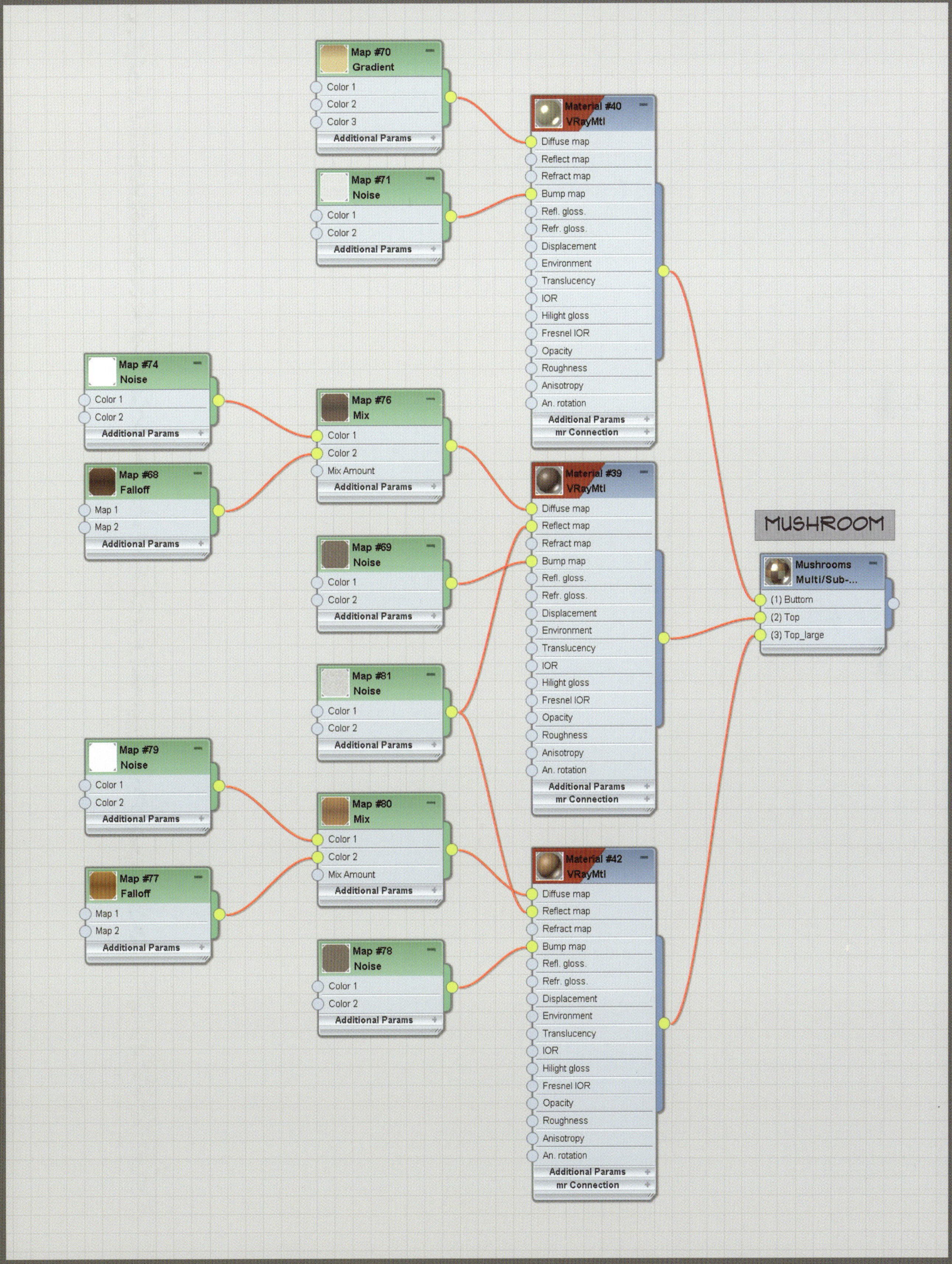

| TABLE-WARE 040 | 3DS MAX 2010 + FBX (V-RAY) | 3,596 KB | CLASSICAL |

MODIFIERS STACK
- NOT COLLAPSED
- COLLAPSED

TEXTURE
- NOT INCLUDED
- INCLUDED

MAPPING
- UNWRAP
- UVW MAP

TOTAL
- POLYS : 323,226
- VERTS : 192,821

TABLE_WARE

MAX

3DS

FBX - SIMPLE OBJECT WITHOUT MATERIALS (WITH MAPPING AND TEXTURES INCLUDED)

*.MAX - V-RAY 1.5 - OR HIGHER - WITH TEXTURES AND SHADERS

V-RAY - OBJECT PREPARED FOR V-RAY RENDERER (WITH TEXTURES AND SHADERS)

*. MAX 2010 - OR HIGHER

ELEMENT OF MODEL

PARTIAL GROUP

1. TABLE_WARE-A   2. TABLE_WARE-B   3. TABLE_WARE-C   4. TABLE_WARE-D

TABLE-WARE VOL I MADE WITH 3DS MAX AND V-RAY

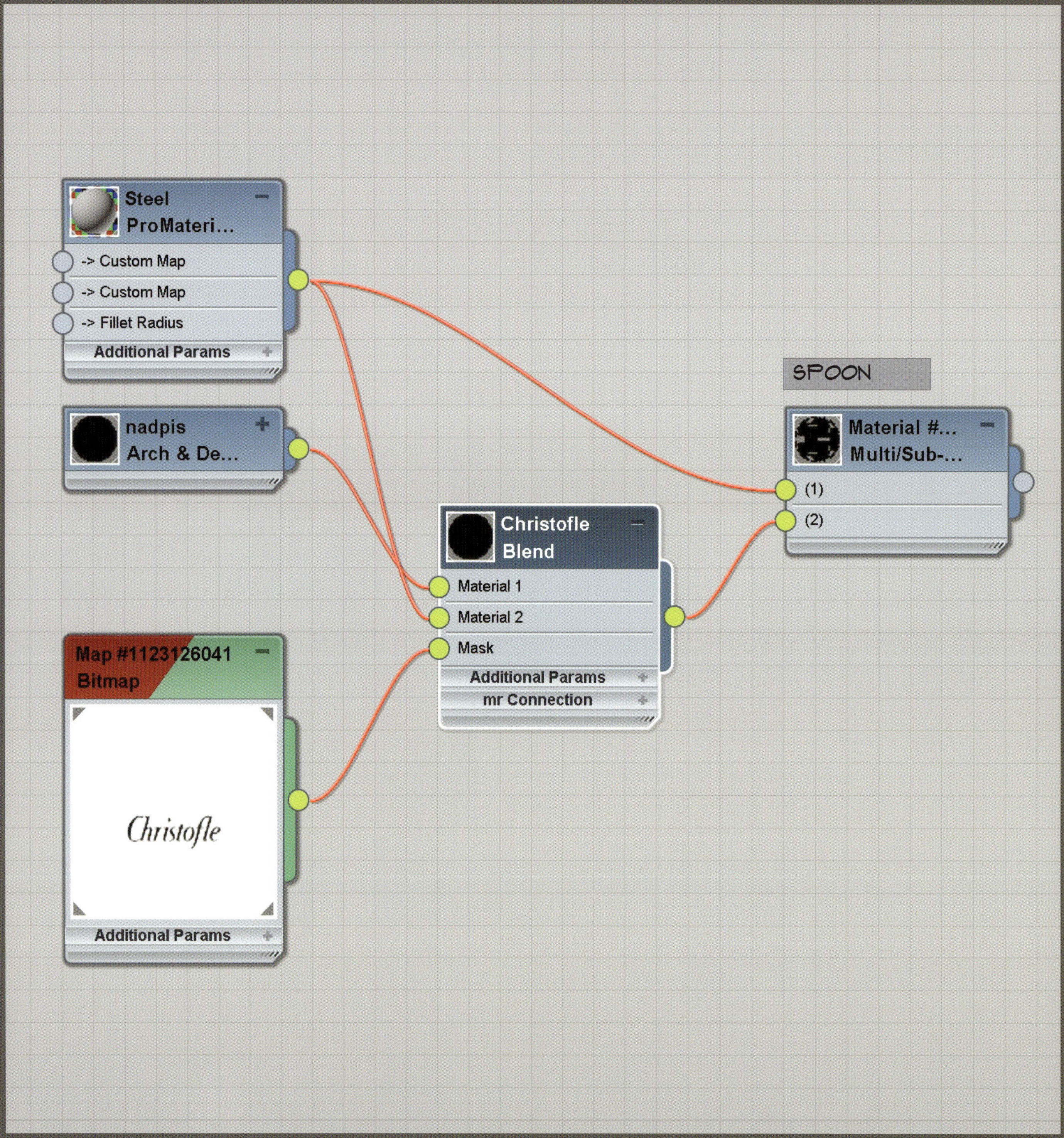

MADE WITH:
THE NEW DIGITAL AGE

SLATE MATERIAL EDITOR

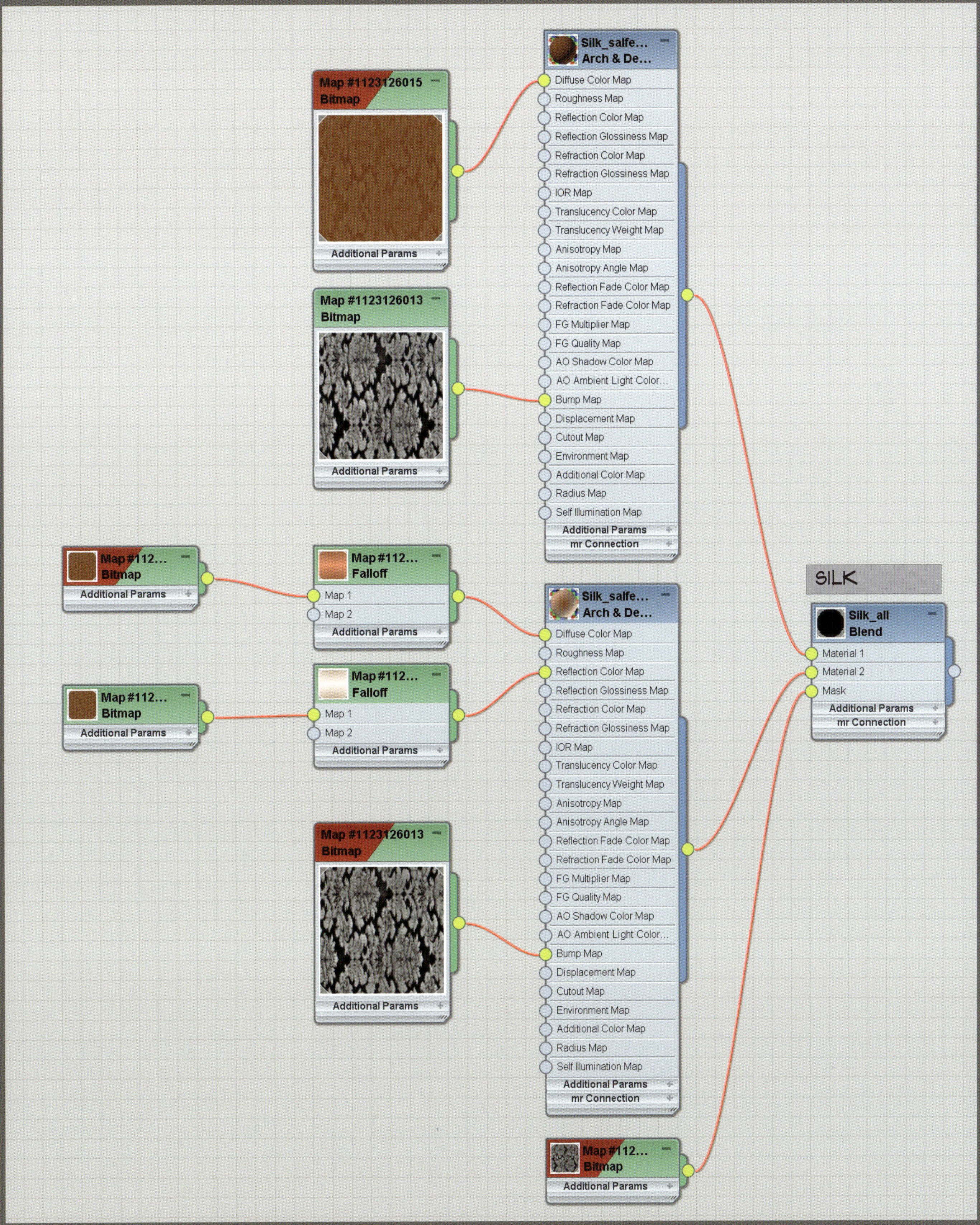

TABLE-WARE VOL I MADE WITH 3DS MAX AND V-RAY

TABLE-WARE 041

MODIFIERS STACK
- NOT COLLAPSED
- COLLAPSED

TEXTURE
- NOT INCLUDED
- INCLUDED

MAPPING
- UNWRAP
- UVW MAP

TOTAL
- POLYS : 736,261
- VERTS : 376,651

3DS MAX 2010 + FBX (V-RAY)   3,184 KB   CLASSICAL

TABLE_WARE

MAX
3DS
FBX - SIMPLE OBJECT WITHOUT MATERIALS (WITH MAPPING AND TEXTURES INCLUDED)
*.MAX - V-RAY 1.5 - OR HIGHER - WITH TEXTURES AND SHADERS
V-RAY - OBJECT PREPARED FOR V-RAY RENDERER (WITH TEXTURES AND SHADERS)

*. MAX 2010 - OR HIGHER

ELEMENT OF MODEL

PARTIAL GROUP

| 1. TABLE_WARE-A | 2. TABLE_WARE-B | 3. TABLE_WARE-C | 4. TABLE_WARE-D |

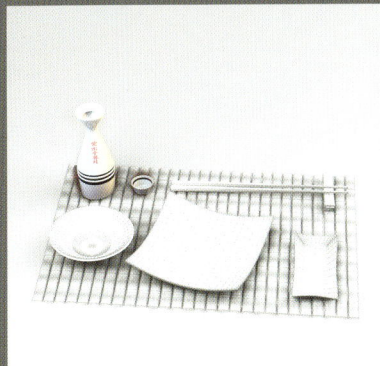

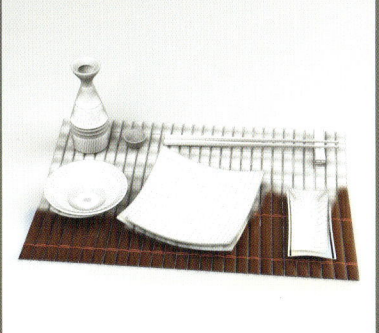

TABLE-WARE VOL 1 MADE WITH 3DS MAX AND V-RAY

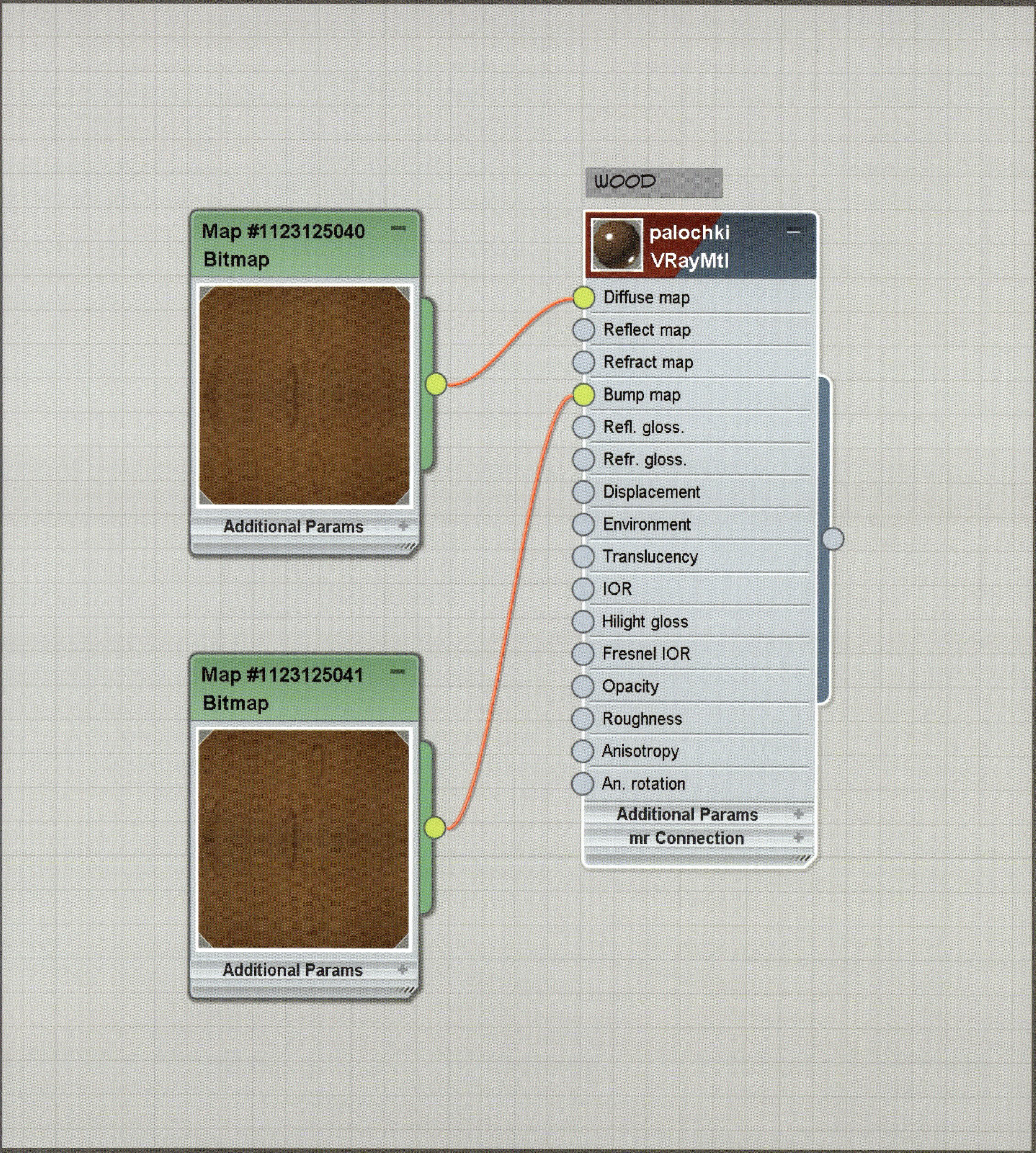

SLATE MATERIAL EDITOR

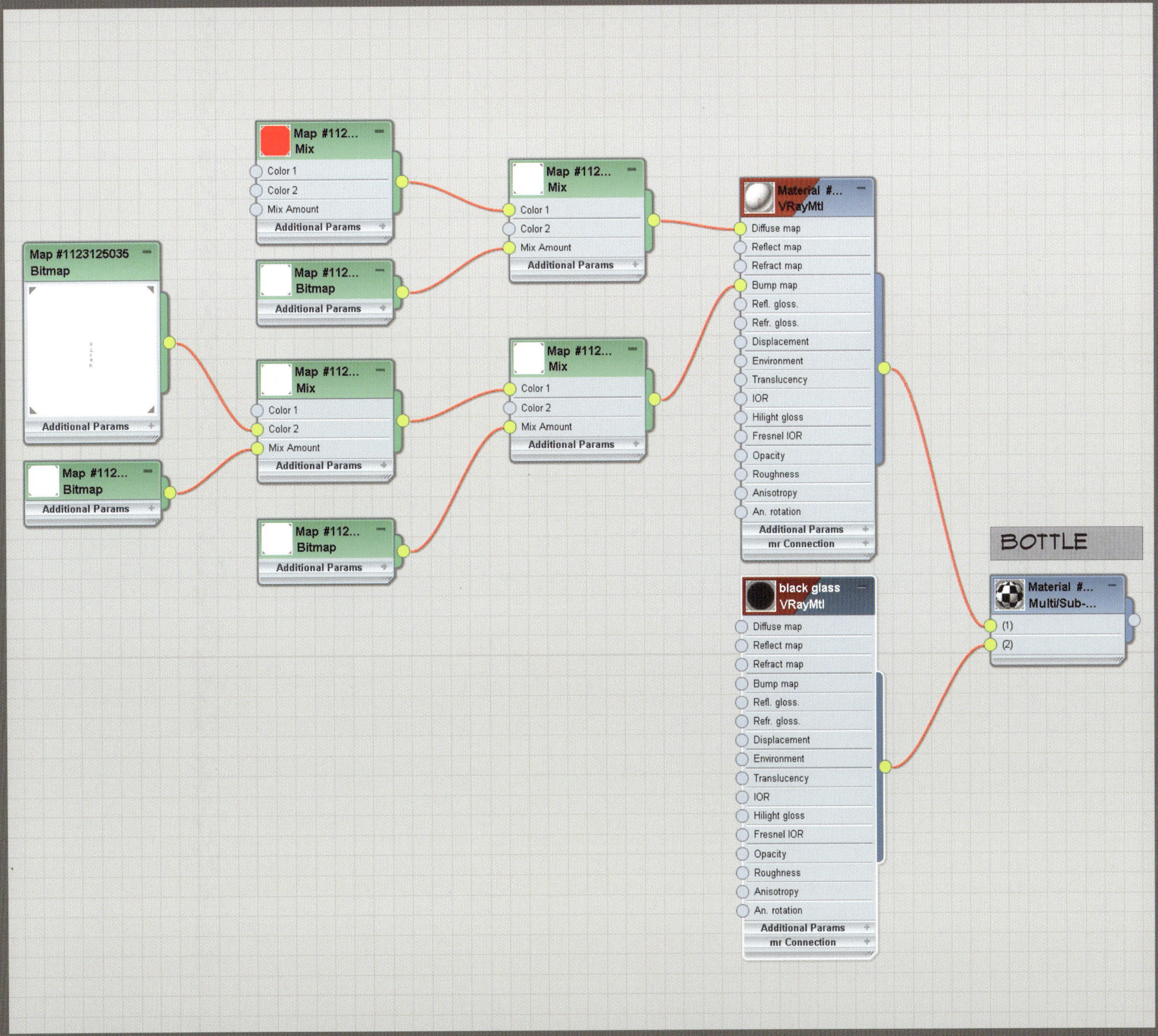

TABLE-WARE VOL I MADE WITH 3DS MAX AND V-RAY

| TABLE-WARE 042 | ⚙ 3DS MAX 2010 + FBX (V-RAY) | 🗄 37.332 KB | ☆ CLASSICAL |

MODIFIERS STACK
NOT COLLAPSED
COLLAPSED

TEXTURE
NOT INCLUDED
INCLUDED

MAPPING
UNWRAP
UVW MAP

TOTAL
POLYS : 242,404
VERTS : 239,786

TABLE_WARE

MAX
3DS
FBX – SIMPLE OBJECT WITHOUT MATERIALS (WITH MAPPING AND TEXTURES INCLUDED)
*.MAX – V-RAY 1.5 – OR HIGHER – WITH TEXTURES AND SHADERS
V-RAY – OBJECT PREPARED FOR V-RAY RENDERER (WITH TEXTURES AND SHADERS)

*. MAX 2010 – OR HIGHER

ELEMENT OF MODEL

PARTIAL GROUP

1. TABLE_WARE-A
2. TABLE_WARE-B
3. TABLE_WARE-C
4. TABLE_WARE-D

TABLE-WARE VOL 1 MADE WITH 3DS MAX AND V-RAY

SLATE MATERIAL EDITOR

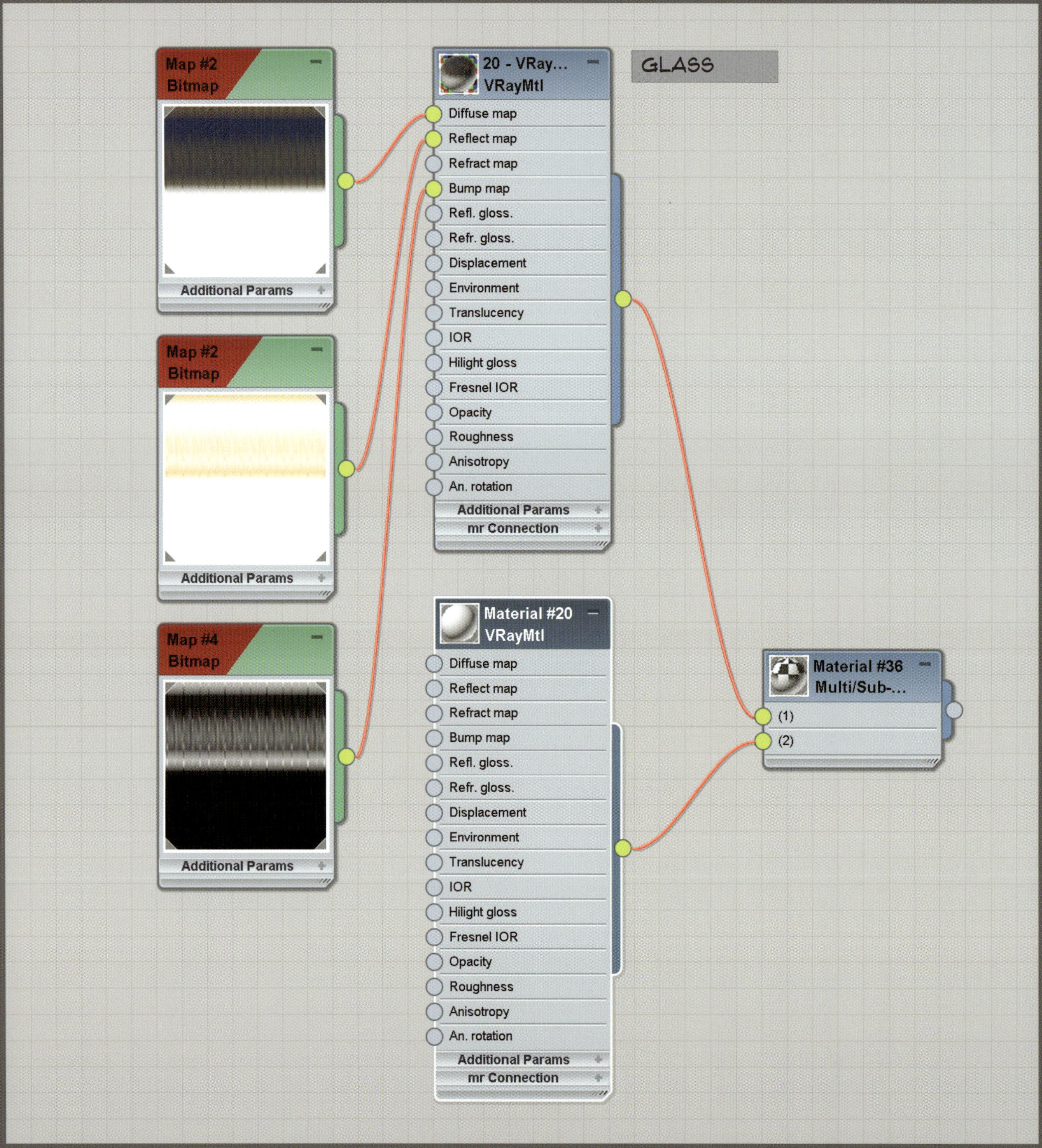

SLATE MATERIAL EDITOR

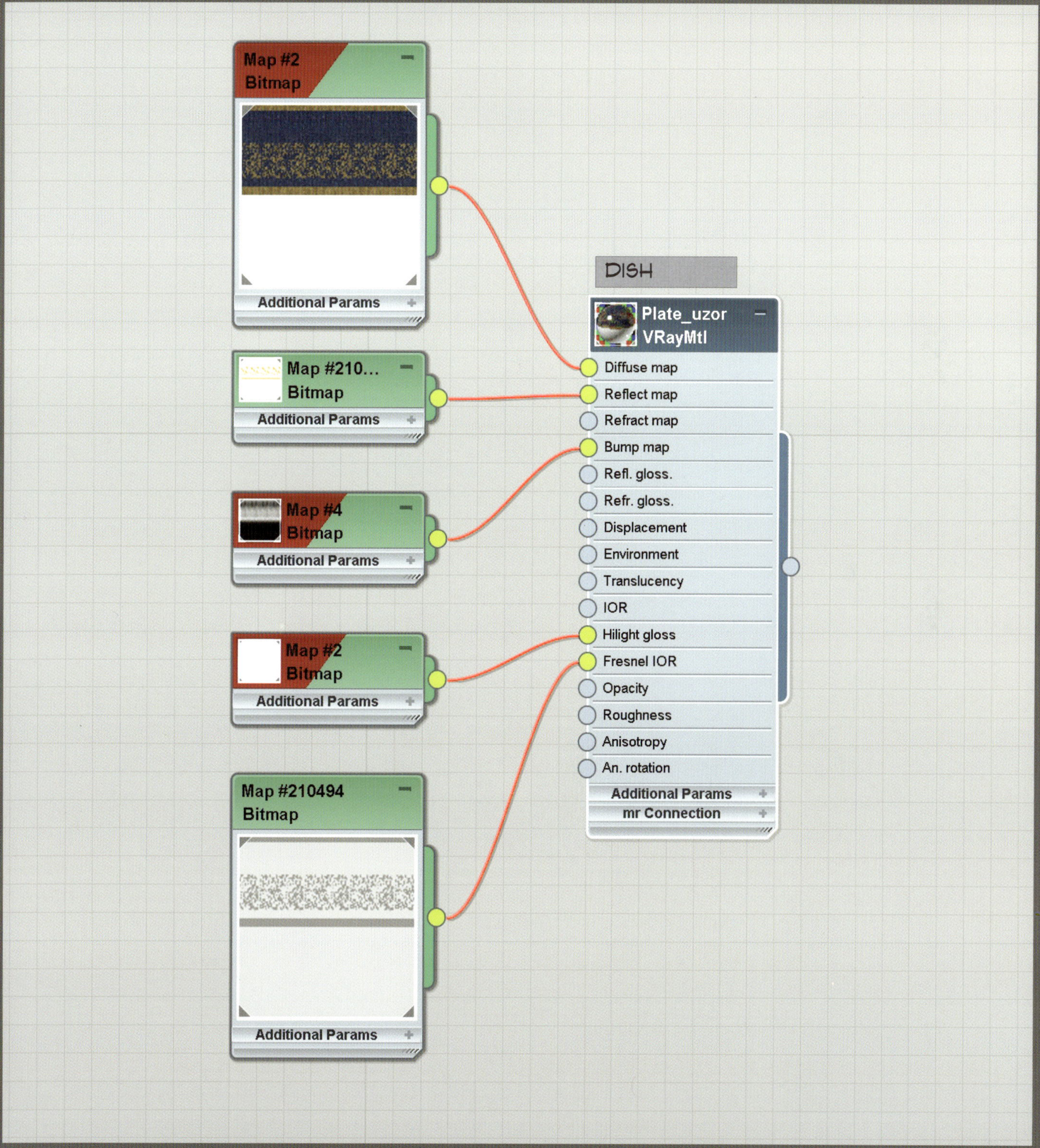

DISH

Plate_uzor — VRayMtl

TABLE-WARE 043

3DS MAX 2010 + FBX (V-RAY) 26,208 KB CLASSICAL

TABLE_WARE

MODIFIERS STACK
- NOT COLLAPSED
- COLLAPSED

TEXTURE
- NOT INCLUDED
- INCLUDED

MAPPING
- UNWRAP
- UVW MAP

TOTAL
- POLYS : 383,378
- VERTS : 218,690

MAX

3DS

FBX – SIMPLE OBJECT WITHOUT MATERIALS (WITH MAPPING AND TEXTURES INCLUDED)

*.MAX – V-RAY 1.5 – OR HIGHER – WITH TEXTURES AND SHADERS

V-RAY – OBJECT PREPARED FOR V-RAY RENDERER (WITH TEXTURES AND SHADERS)

*. MAX 2010 – OR HIGHER

ELEMENT OF MODEL

PARTIAL GROUP

1. TABLE_WARE-A   2. TABLE_WARE-B   3. TABLE_WARE-C   4. TABLE_WARE-D

TABLE-WARE VOL 1 MADE WITH 3DS MAX AND V-RAY

SLATE MATERIAL EDITOR

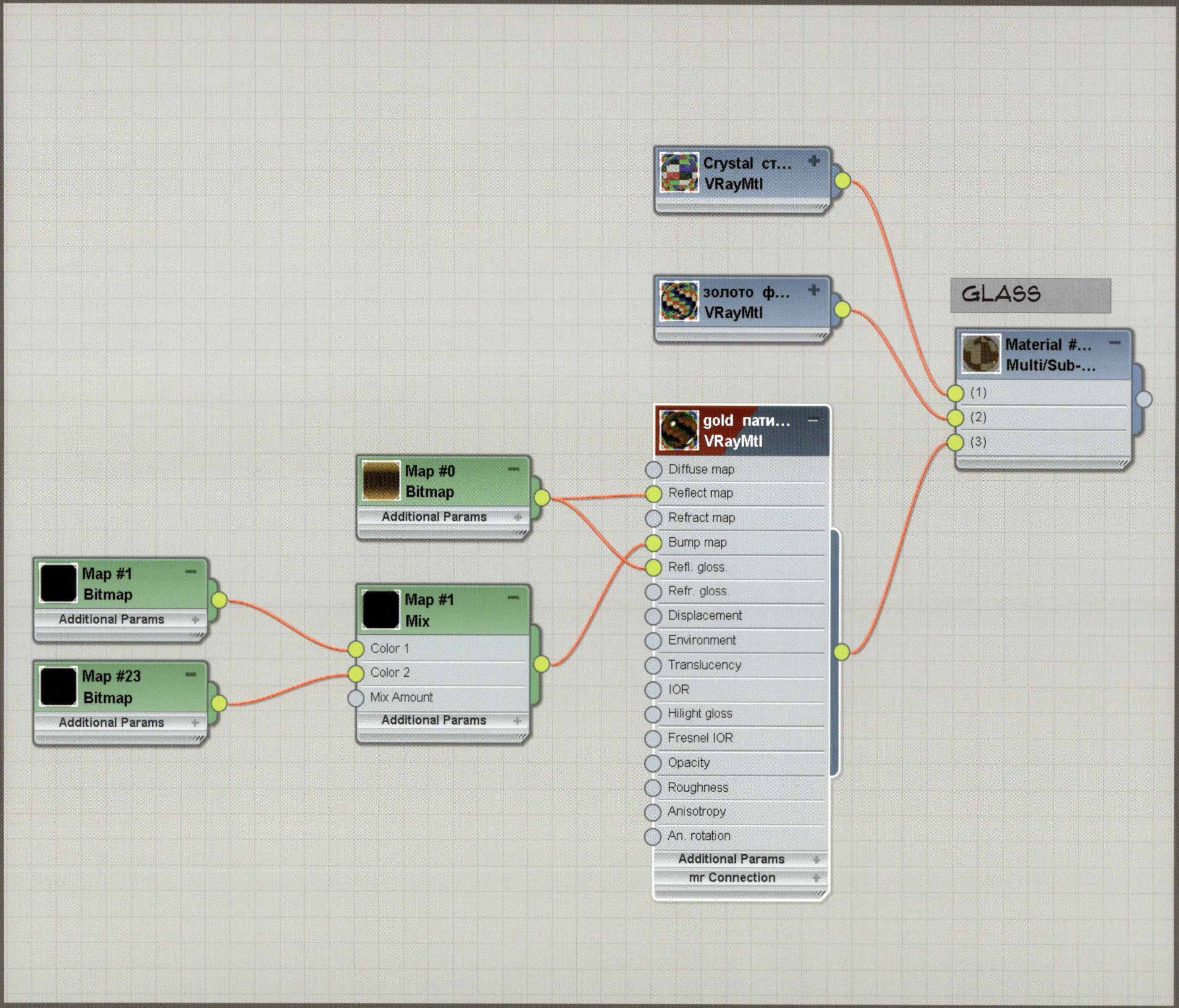

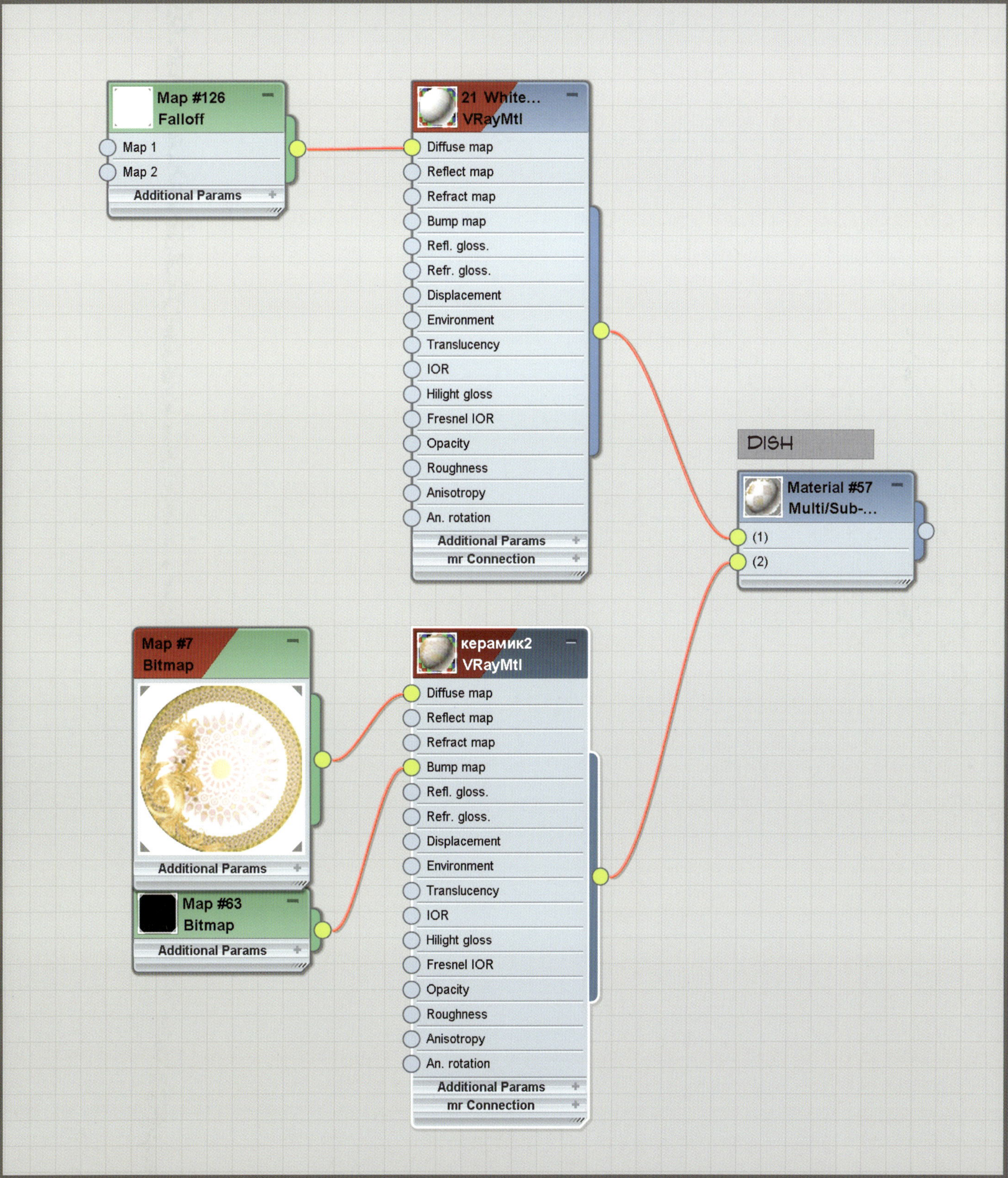

TABLE-WARE 044

3DS MAX 2010 + FBX (V-RAY)   1,532 KB   CLASSICAL

MODIFIERS STACK
NOT COLLAPSED
COLLAPSED

TEXTURE
NOT INCLUDED
INCLUDED

MAPPING
UNWRAP
UVW MAP

TOTAL
POLYS : 19,036
VERTS : 20,069

TABLE_WARE

MAX

3DS

FBX – SIMPLE OBJECT WITHOUT MATERIALS (WITH MAPPING AND TEXTURES INCLUDED)

*.MAX – V-RAY 1.5 – OR HIGHER – WITH TEXTURES AND SHADERS

V-RAY – OBJECT PREPARED FOR V-RAY RENDERER (WITH TEXTURES AND SHADERS)

*. MAX 2010 – OR HIGHER

ELEMENT OF MODEL

PARTIAL GROUP

1. TABLE_WARE-A
2. TABLE_WARE-B
3. TABLE_WARE-C
4. TABLE_WARE-D

TABLE-WARE VOL I MADE WITH 3DS MAX AND V-RAY

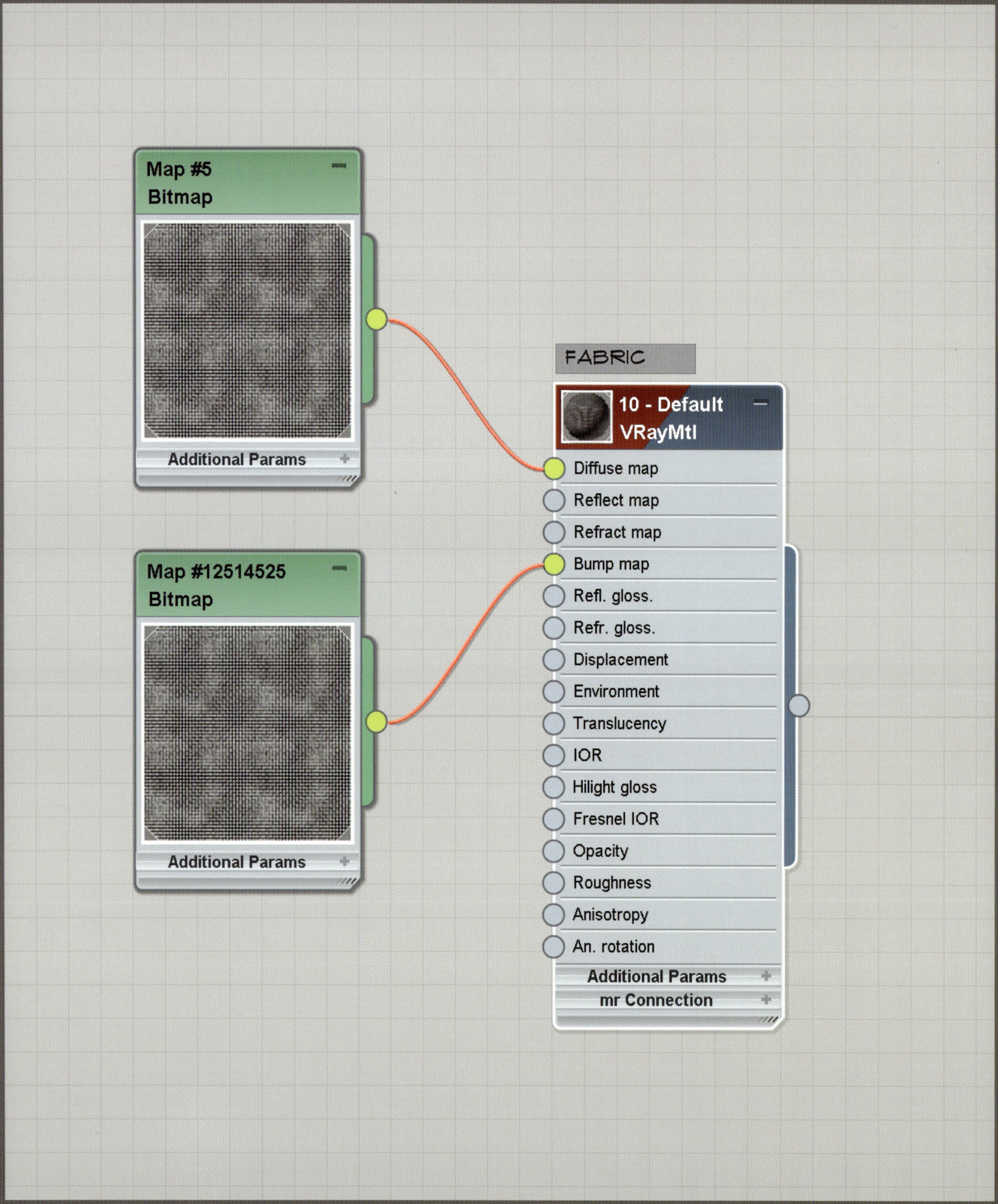

SLATE MATERIAL EDITOR

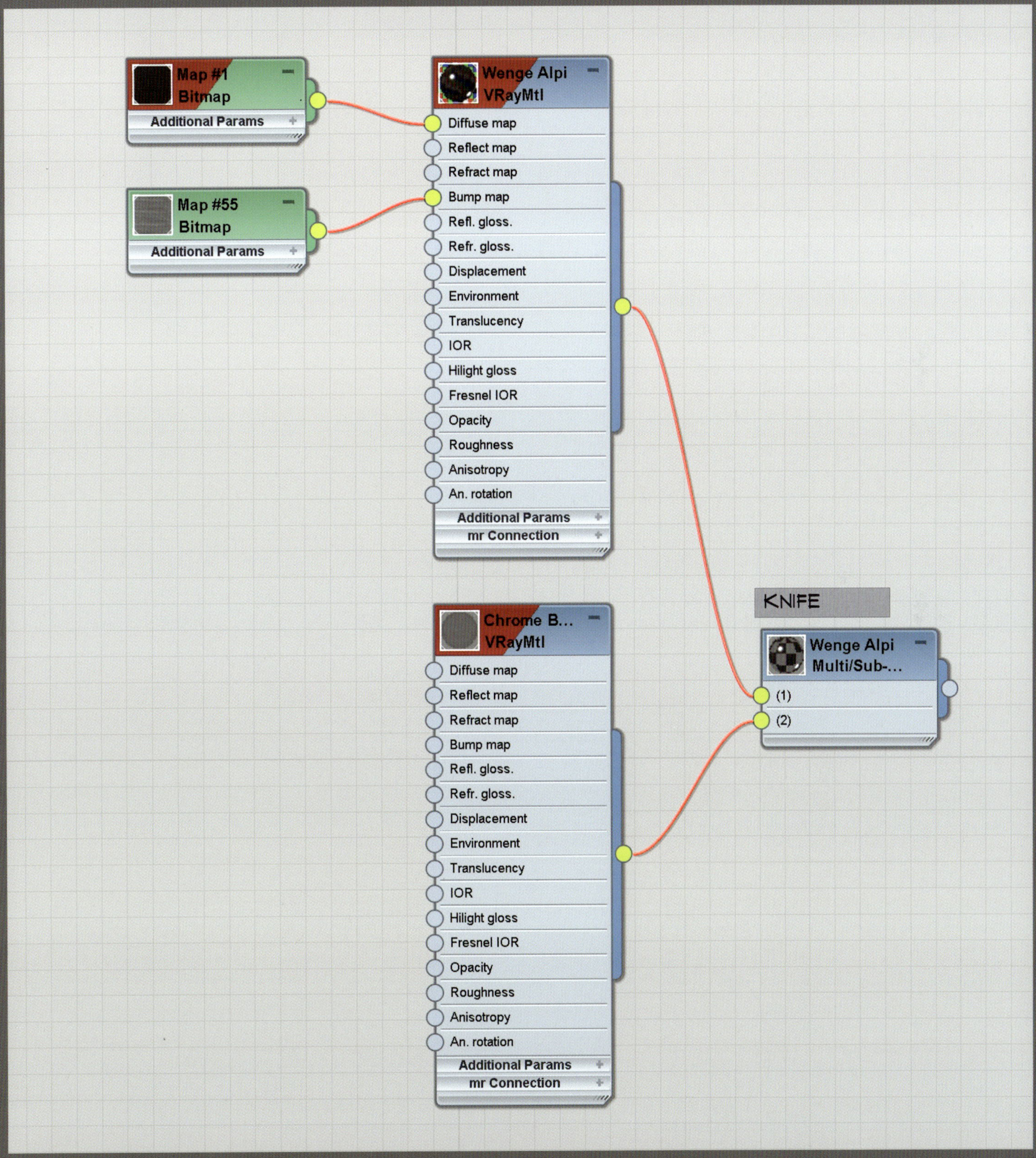

KNIFE

TABLE-WARE VOL I MADE WITH 3DS MAX AND V-RAY

TABLE-WARE 045

3DS MAX 2010 + FBX (V-RAY)   12,784 KB   CLASSICAL

MODIFIERS STACK
NOT COLLAPSED
COLLAPSED

TEXTURE
NOT INCLUDED
INCLUDED

MAPPING
UNWRAP
UVW MAP

TOTAL
POLYS : 47,295
VERTS : 50,300

TABLE_WARE

MAX

3DS

FBX – SIMPLE OBJECT WITHOUT MATERIALS (WITH MAPPING AND TEXTURES INCLUDED)

*.MAX – V-RAY 1.5 – OR HIGHER – WITH TEXTURES AND SHADERS

V-RAY – OBJECT PREPARED FOR V-RAY RENDERER (WITH TEXTURES AND SHADERS)

*. MAX 2010 – OR HIGHER

MADE WITH:
THE NEW DIGITAL AGE

ELEMENT OF MODEL

PARTIAL GROUP

1. TABLE_WARE-A   2. TABLE_WARE-B   3. TABLE_WARE-C   4. TABLE_WARE-D

TABLE-WARE VOL 1 MADE WITH 3DS MAX AND V-RAY

SLATE MATERIAL EDITOR

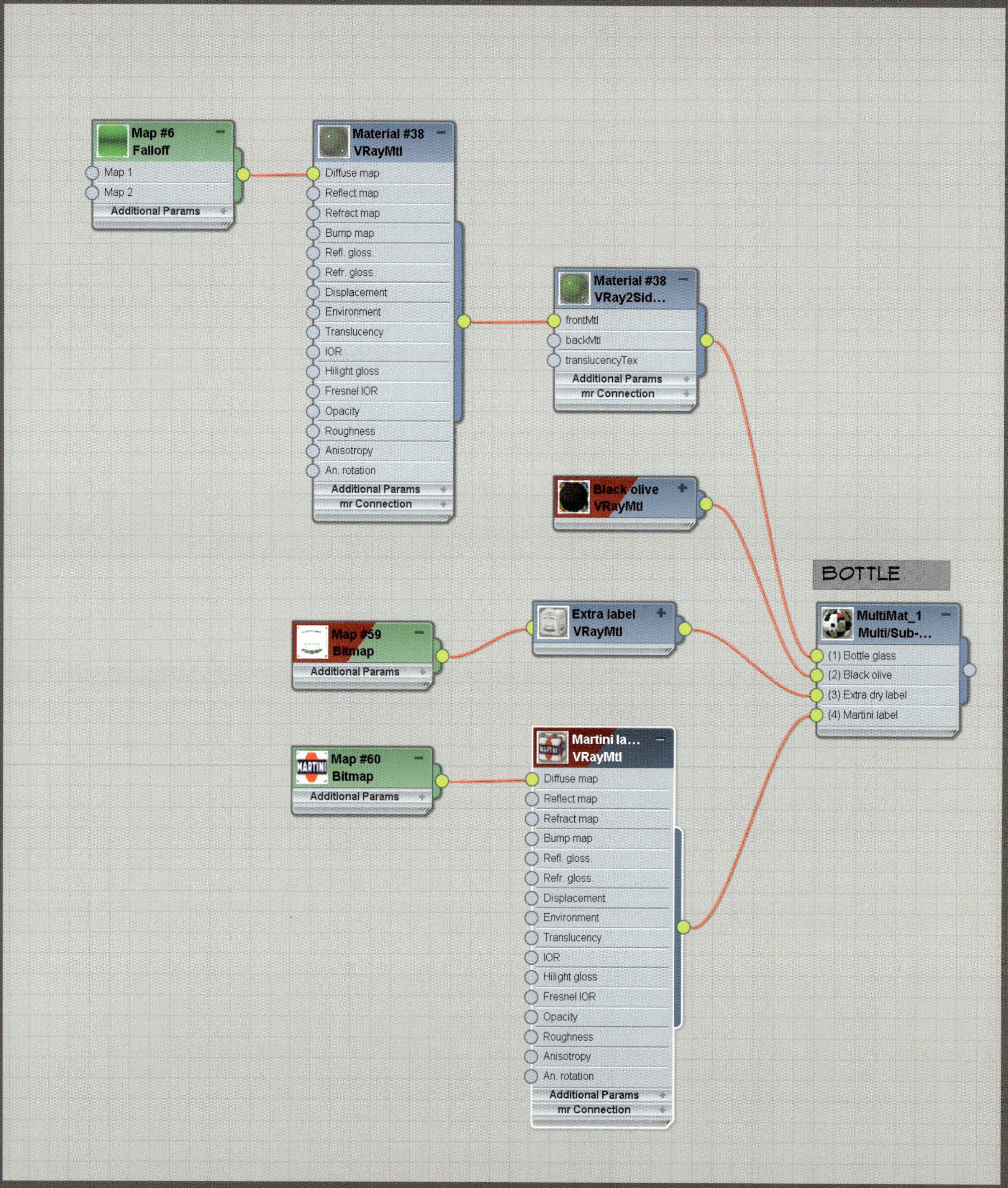

SLATE MATERIAL EDITOR

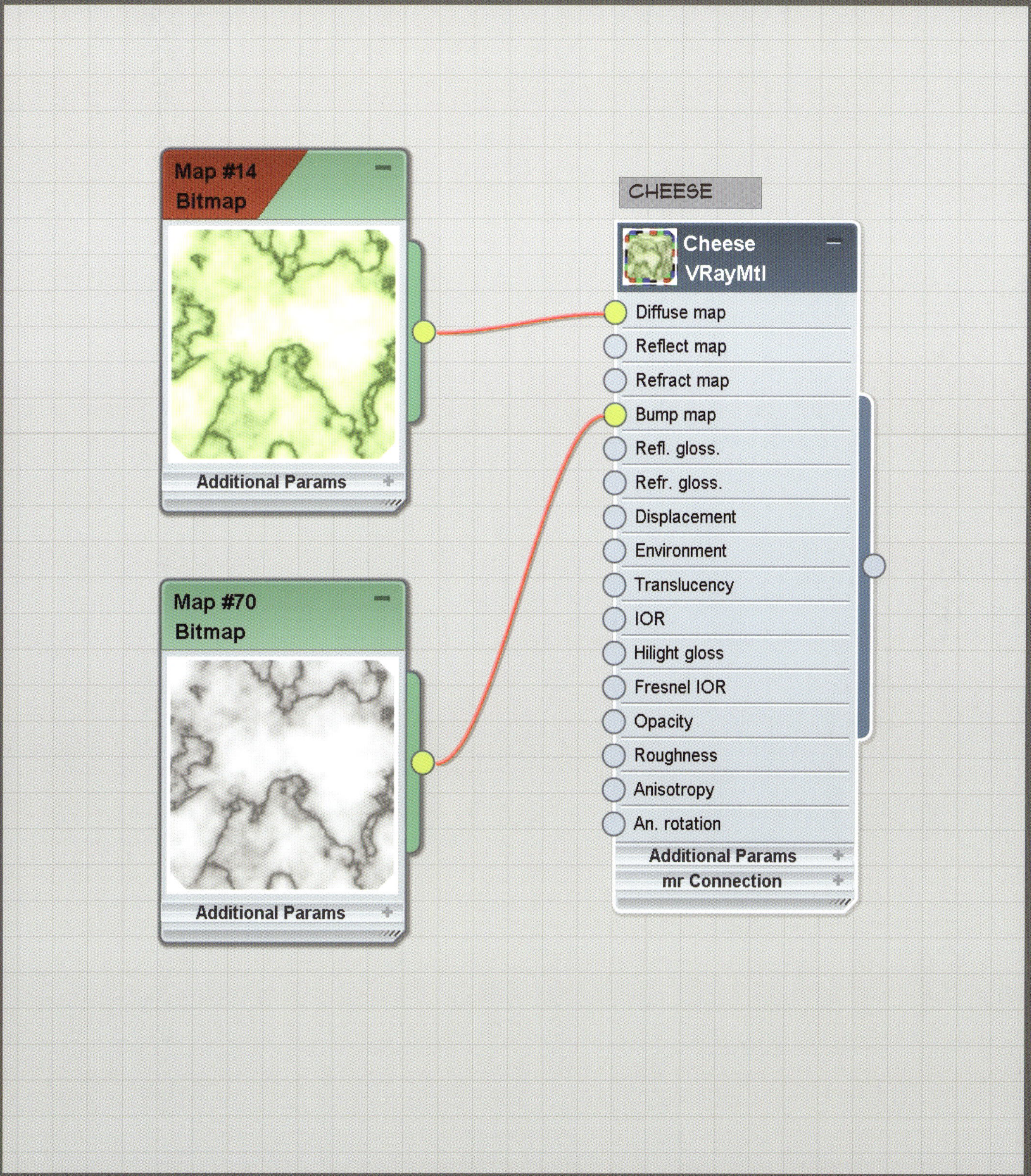

TABLE-WARE VOL 1 MADE WITH 3DS MAX AND V-RAY

TABLE-WARE 046

MODIFIERS STACK
- NOT COLLAPSED
- COLLAPSED

TEXTURE
- NOT INCLUDED
- INCLUDED

MAPPING
- UNWRAP
- UVW MAP

TOTAL
- POLYS : 363,510
- VERTS : 355,122

3DS MAX 2010 + FBX (V-RAY)   8,796 KB   CLASSICAL

TABLE_WARE

MAX

3DS

FBX – SIMPLE OBJECT WITHOUT MATERIALS (WITH MAPPING AND TEXTURES INCLUDED)

*.MAX – V-RAY 1.5 – OR HIGHER – WITH TEXTURES AND SHADERS

V-RAY – OBJECT PREPARED FOR V-RAY RENDERER (WITH TEXTURES AND SHADERS)

*. MAX 2010 – OR HIGHER

ELEMENT OF MODEL

PARTIAL GROUP

1. TABLE_WARE-A   2. TABLE_WARE-B   3. TABLE_WARE-C   4. TABLE_WARE-D

TABLE-WARE VOL 1 MADE WITH 3DS MAX AND V-RAY

SLATE MATERIAL EDITOR

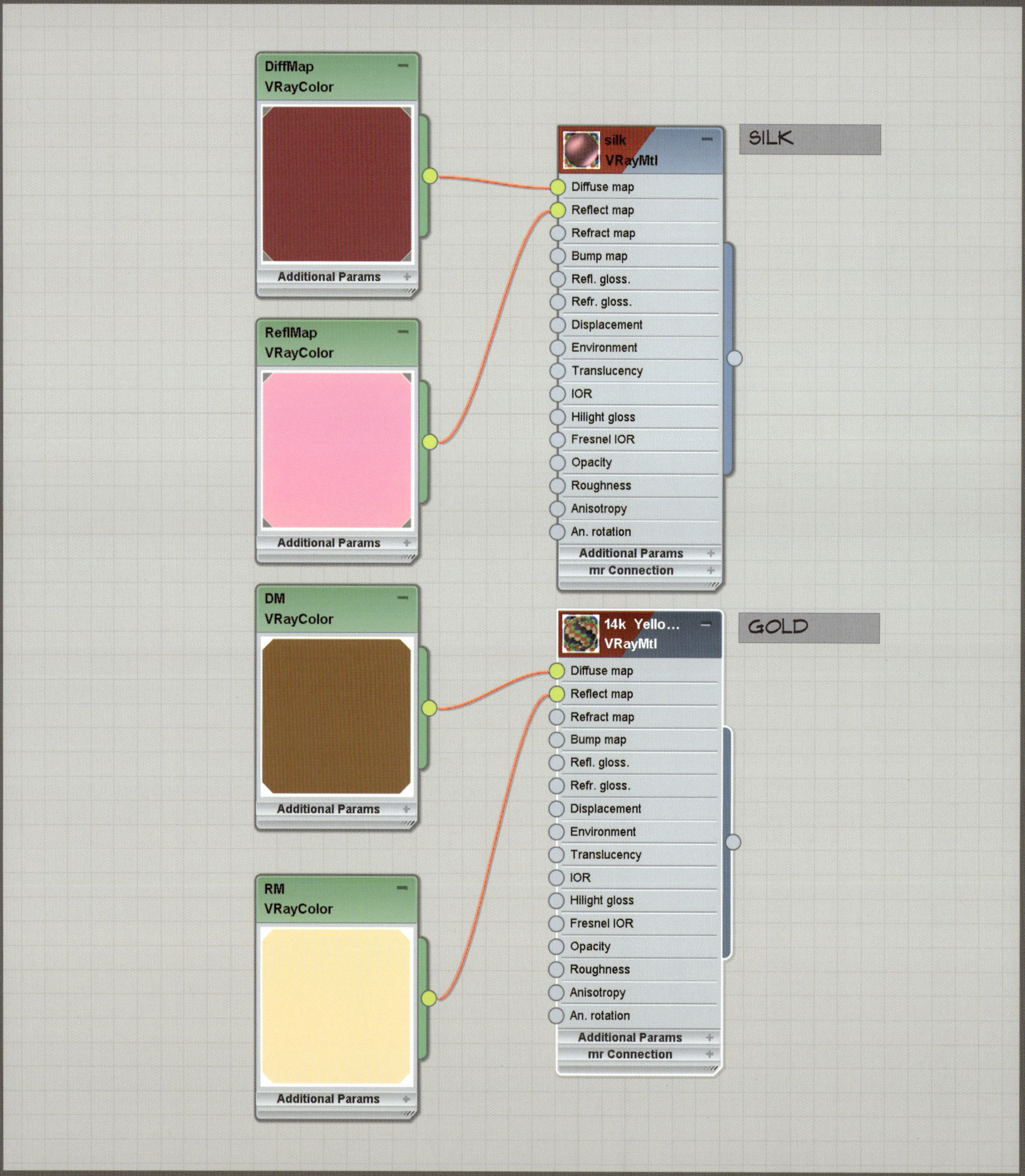

SILK

GOLD

SLATE MATERIAL EDITOR

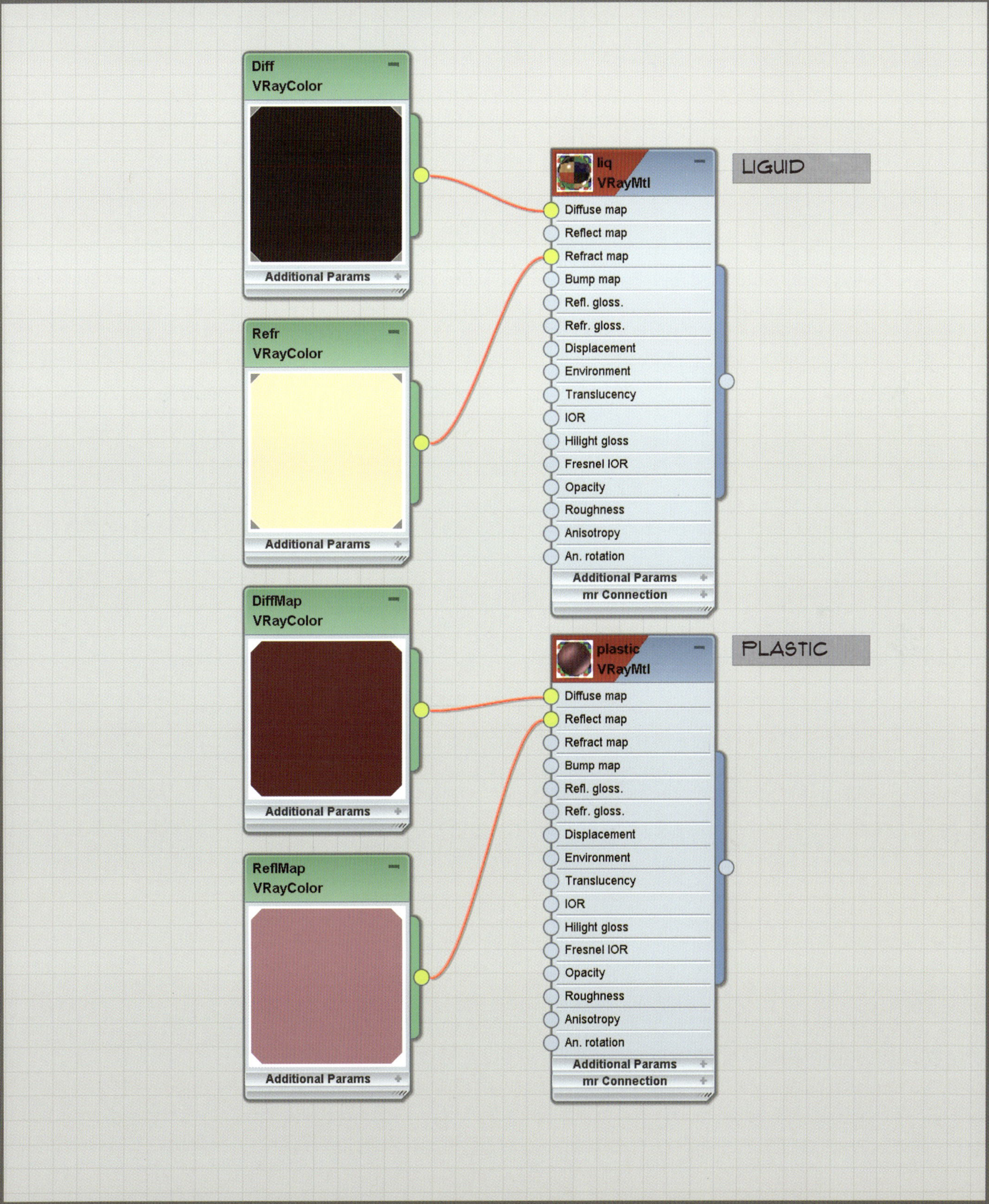

TABLE-WARE VOL 1 MADE WITH 3DS MAX AND V-RAY

TABLE-WARE 041

3DS MAX 2010 + FBX (V-RAY)   100.300 KB   CLASSICAL

MODIFIERS STACK
NOT COLLAPSED
COLLAPSED

TEXTURE
NOT INCLUDED
INCLUDED

MAPPING
UNWRAP
UVW MAP

TOTAL
POLYS : 1,910,975
VERTS : 1,882,264

TABLE_WARE

MAX
3DS
FBX – SIMPLE OBJECT WITHOUT MATERIALS (WITH MAPPING AND TEXTURES INCLUDED)
*.MAX – V-RAY 1.5 – OR HIGHER – WITH TEXTURES AND SHADERS
V-RAY – OBJECT PREPARED FOR V-RAY RENDERER (WITH TEXTURES AND SHADERS)

*. MAX 2010 – OR HIGHER

ELEMENT OF MODEL

PARTIAL GROUP

1. TABLE_WARE-A 2. TABLE_WARE-B 3. TABLE_WARE-C 4. TABLE_WARE-D

TABLE-WARE VOL 1 MADE WITH 3DS MAX AND V-RAY

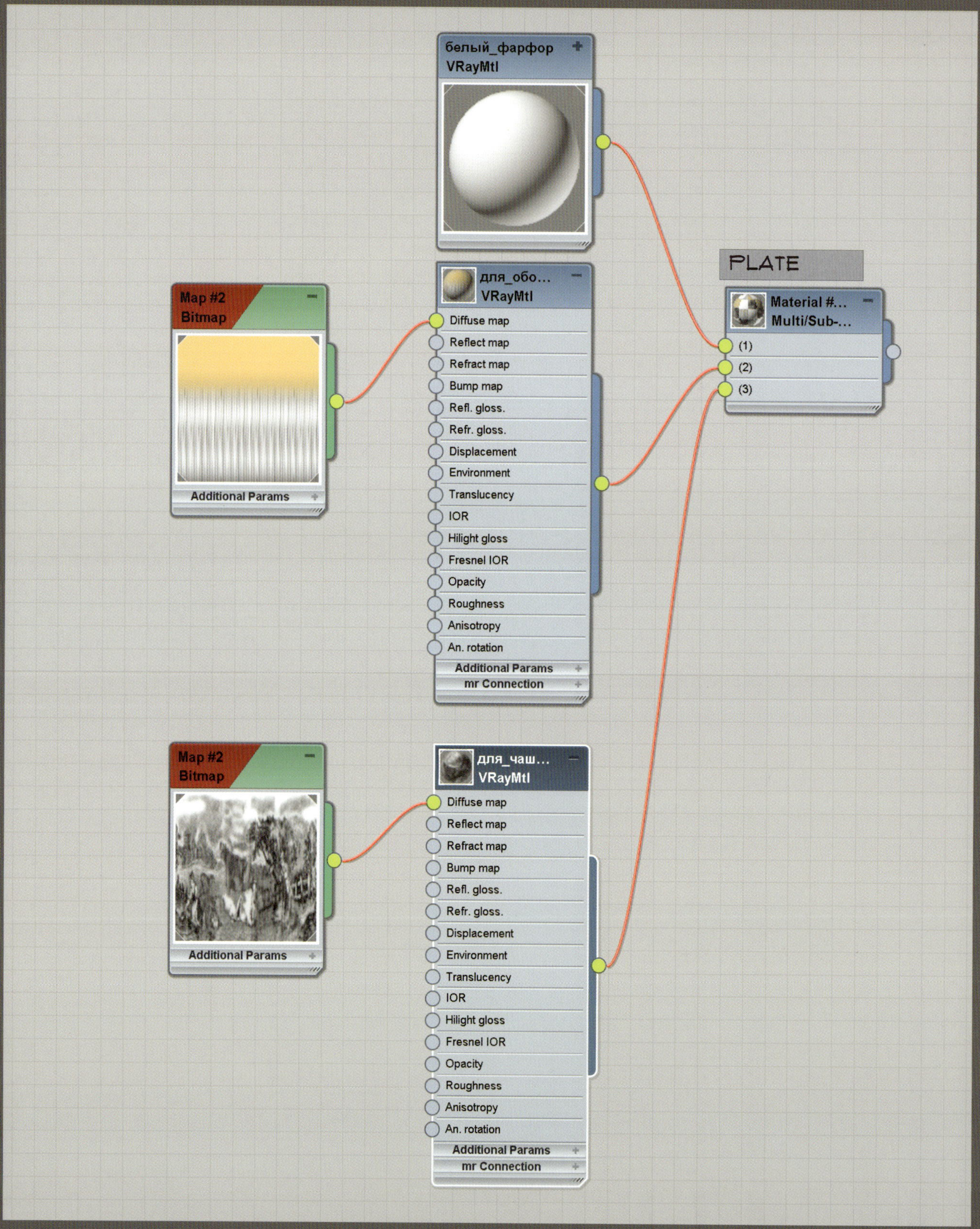

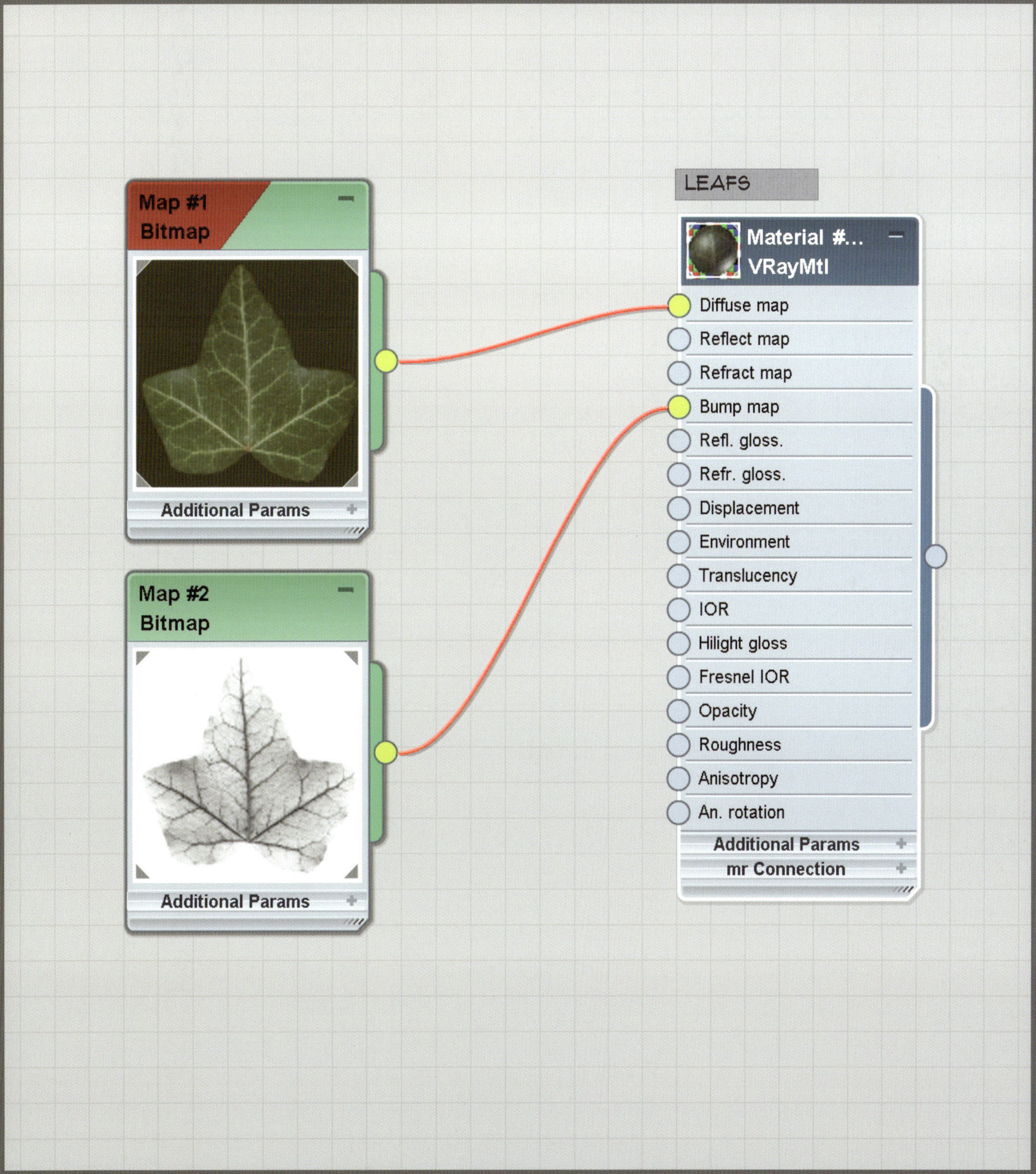

TABLE-WARE 048

3DS MAX 2010 + FBX (V-RAY)   19,136 KB   CLASSICAL

MODIFIERS STACK
- NOT COLLAPSED
- COLLAPSED

TEXTURE
- NOT INCLUDED
- INCLUDED

MAPPING
- UNWRAP
- UVW MAP

TOTAL
- POLYS : 210,549
- VERTS : 199,545

TABLE_WARE

MAX

3DS

FBX – SIMPLE OBJECT WITHOUT MATERIALS (WITH MAPPING AND TEXTURES INCLUDED)

*.MAX – V-RAY 1.5 – OR HIGHER – WITH TEXTURES AND SHADERS

V-RAY – OBJECT PREPARED FOR V-RAY RENDERER (WITH TEXTURES AND SHADERS)

*. MAX 2010 – OR HIGHER

ELEMENT OF MODEL

PARTIAL GROUP

1. TABLE_WARE-A   2. TABLE_WARE-B   3. TABLE_WARE-C   4. TABLE_WARE-D

TABLE-WARE VOL 1 MADE WITH 3DS MAX AND V-RAY

SLATE MATERIAL EDITOR

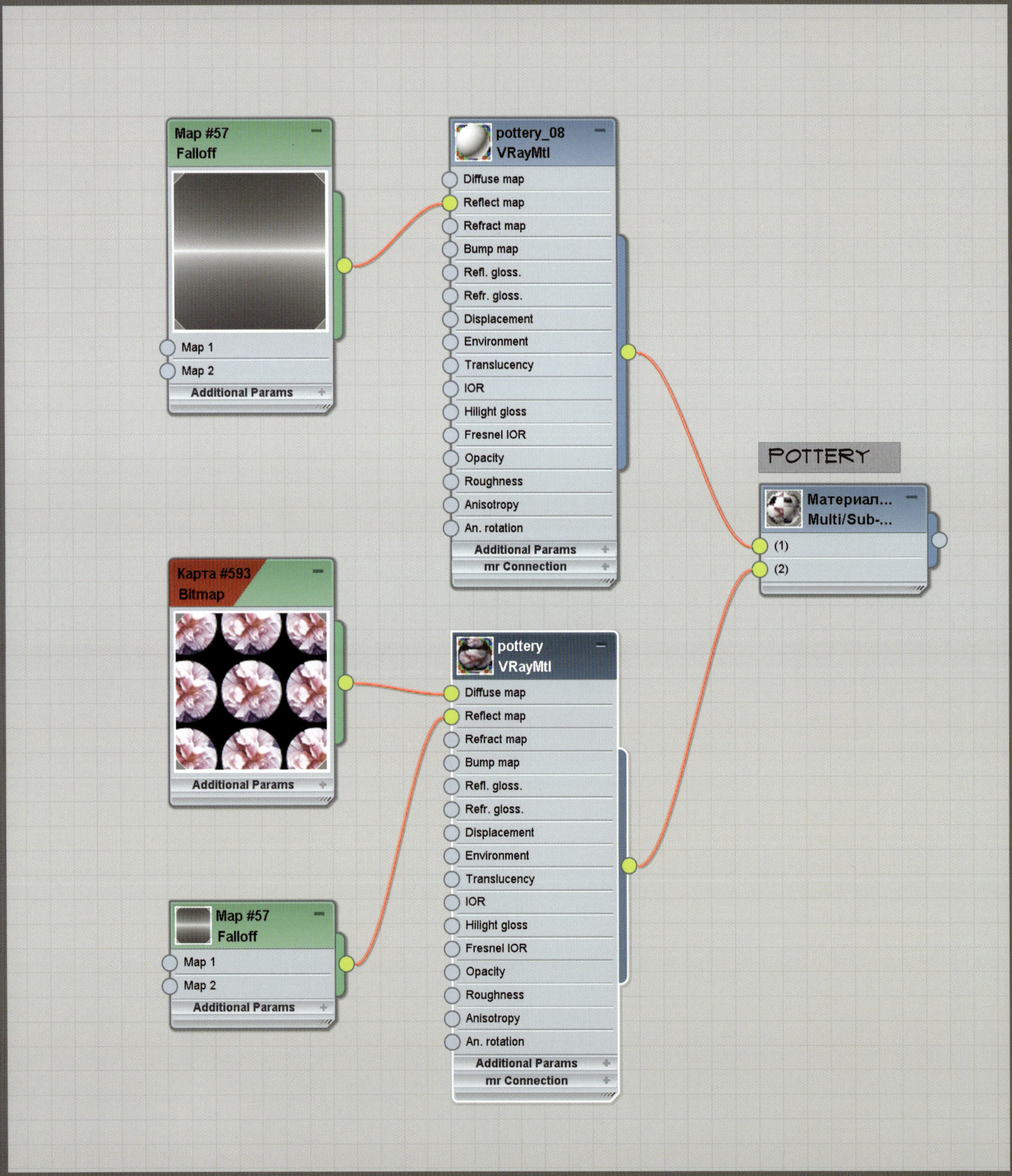

TABLE-WARE VOL 1 MADE WITH 3DS MAX AND V-RAY

SLATE MATERIAL EDITOR

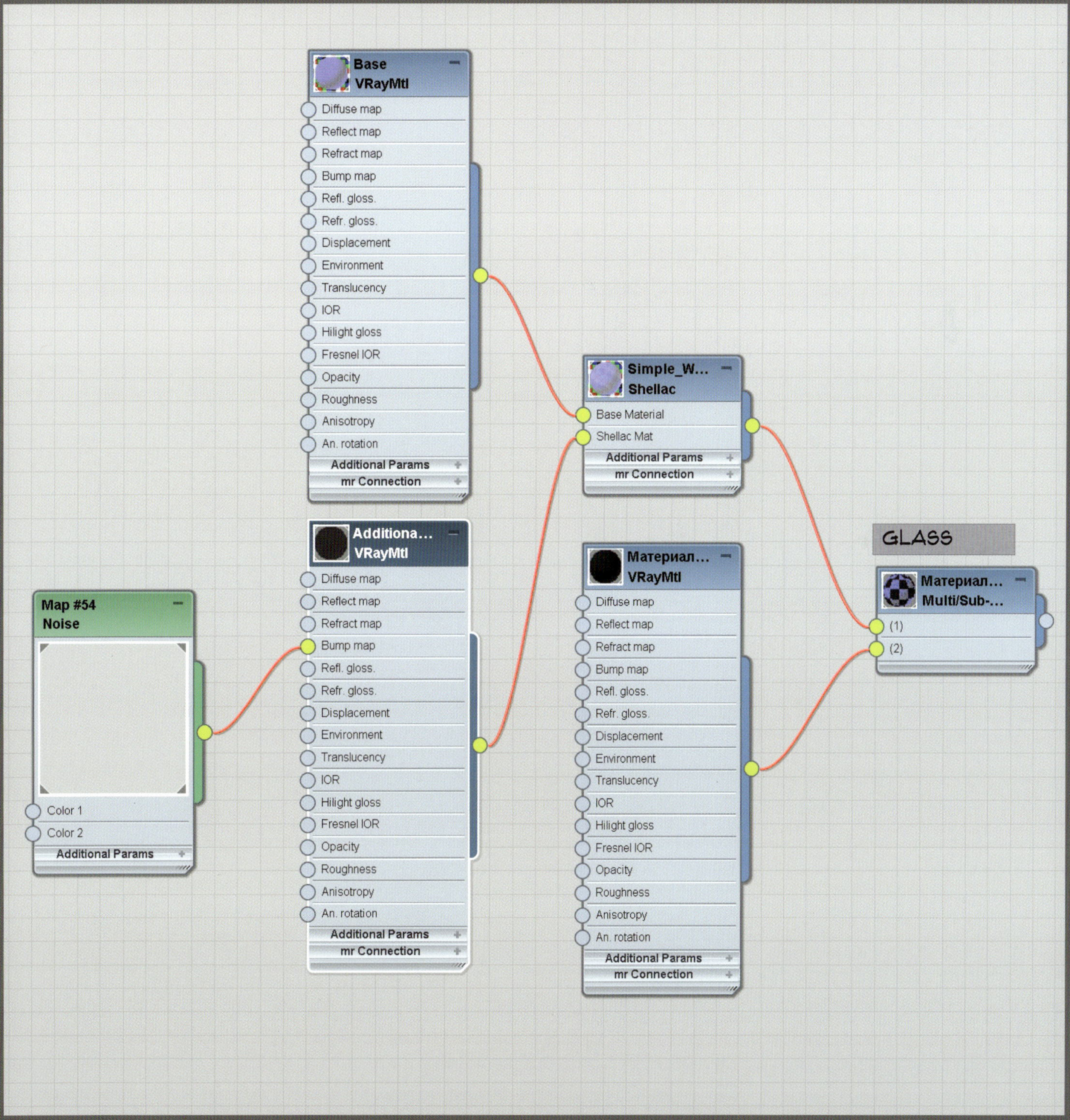

TABLE-WARE VOL I MADE WITH 3DS MAX AND V-RAY

TABLE-WARE 049

3DS MAX 2010 + FBX (V-RAY)   56,864 KB   MODERN

MODIFIERS STACK
- NOT COLLAPSED
- COLLAPSED

TEXTURE
- NOT INCLUDED
- INCLUDED

MAPPING
- UNWRAP
- UVW MAP

TOTAL
- POLYS : 498,050
- VERTS : 426,237

TABLE_WARE

MAX
3DS
FBX – SIMPLE OBJECT WITHOUT MATERIALS (WITH MAPPING AND TEXTURES INCLUDED)
*.MAX – V-RAY 1.5 – OR HIGHER – WITH TEXTURES AND SHADERS
V-RAY – OBJECT PREPARED FOR V-RAY RENDERER (WITH TEXTURES AND SHADERS)

*. MAX 2010 – OR HIGHER

ELEMENT OF MODEL

PARTIAL GROUP

1. TABLE_WARE-A   2. TABLE_WARE-B   3. TABLE_WARE-C   4. TABLE_WARE-D

TABLE-WARE VOL 1 MADE WITH 3DS MAX AND V-RAY

SLATE MATERIAL EDITOR

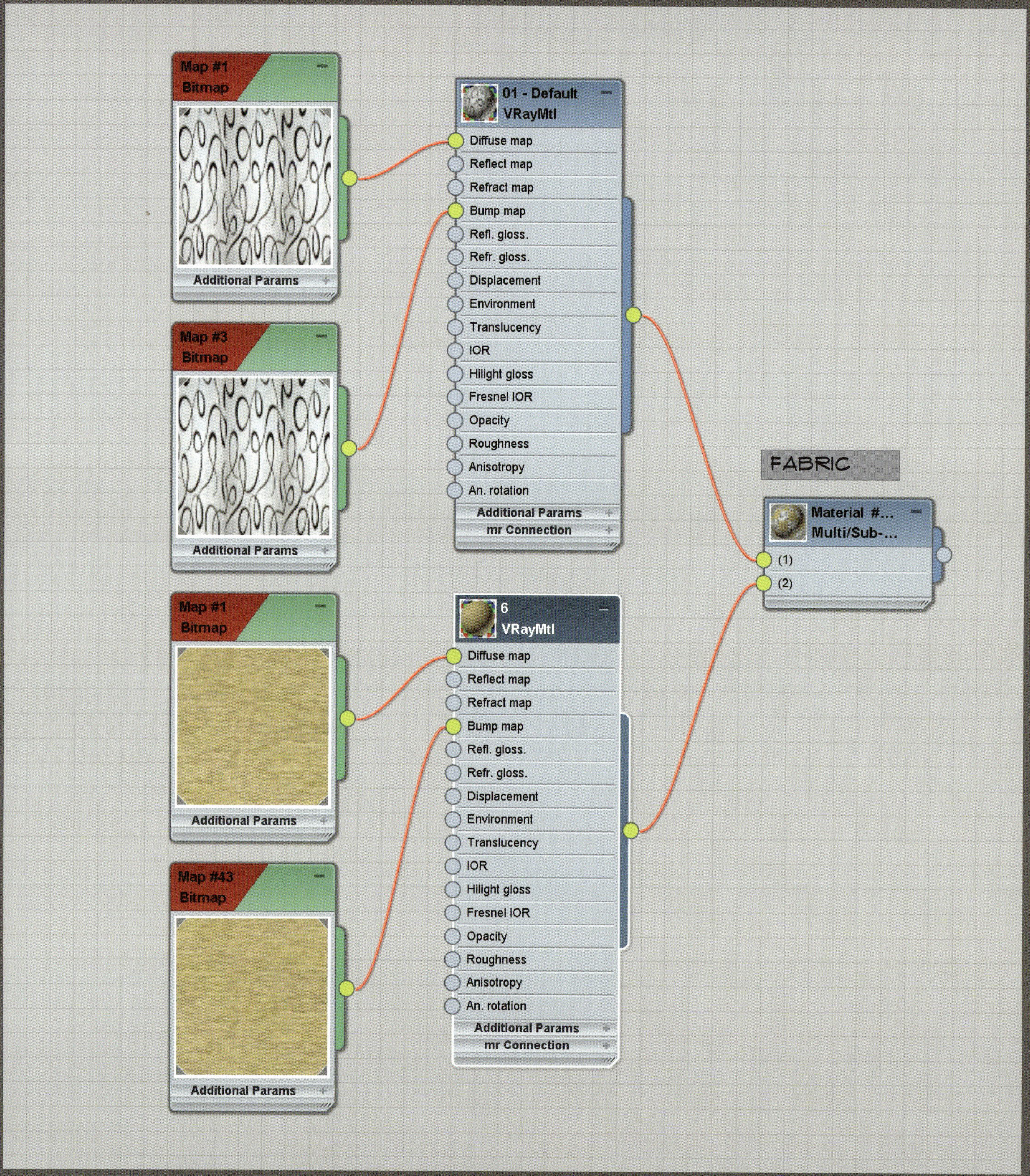

FABRIC

TABLE-WARE VOL I MADE WITH 3DS MAX AND V-RAY

SLATE MATERIAL EDITOR

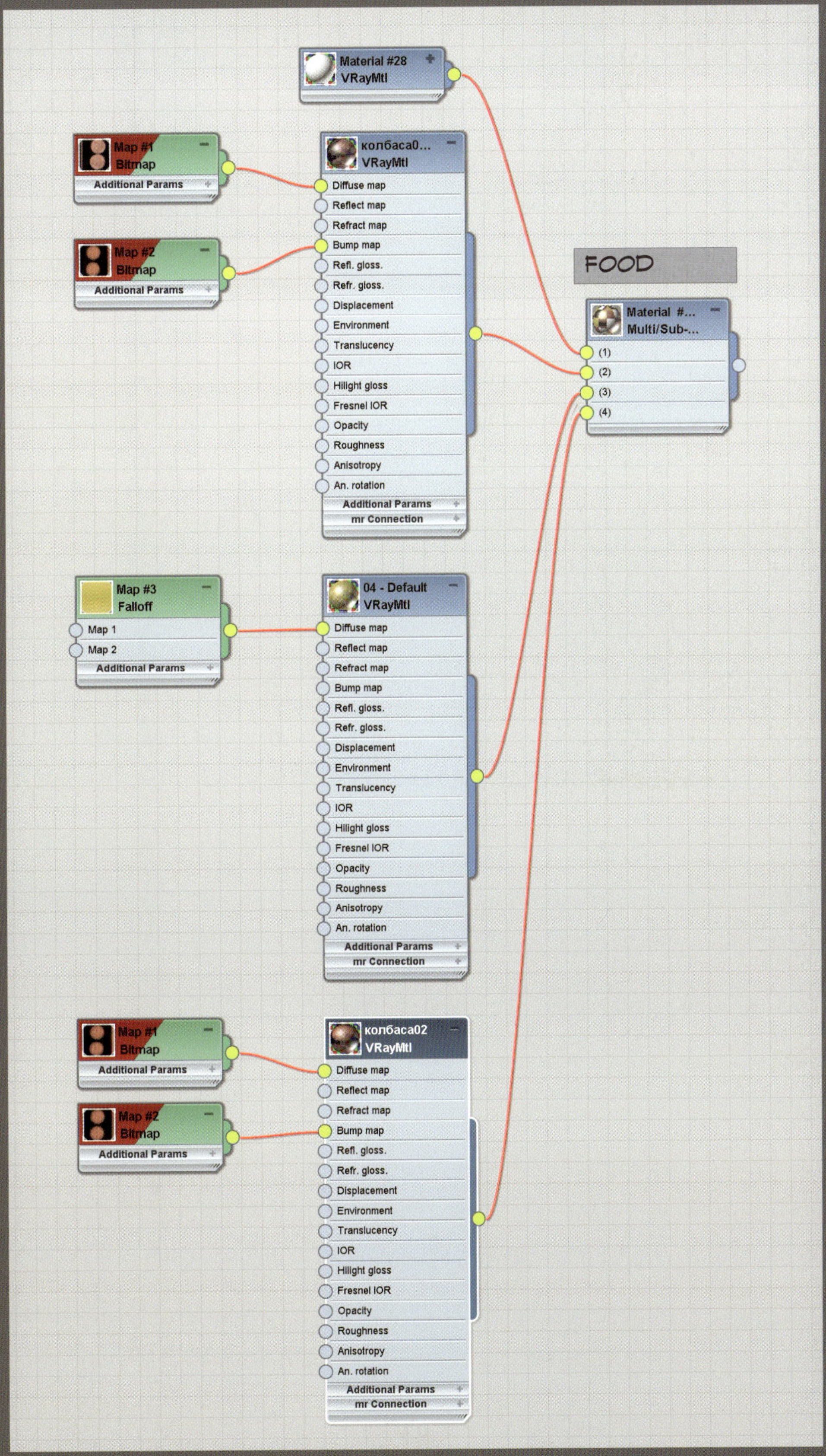

TABLE-WARE VOL I MADE WITH 3DS MAX AND V-RAY

TABLE-WARE 050

3DS MAX 2010 + FBX (V-RAY)   40,992 KB   CLASSICAL

VELLEROY&BOCH

MODIFIERS STACK
NOT COLLAPSED
COLLAPSED

TEXTURE
NOT INCLUDED
INCLUDED

MAPPING
UNWRAP
UVW MAP

TOTAL
POLYS : 307,298
VERTS : 282,815

MAX

3DS

FBX - SIMPLE OBJECT WITHOUT MATERIALS (WITH MAPPING AND TEXTURES INCLUDED)

*.MAX - V-RAY 1.5 - OR HIGHER - WITH TEXTURES AND SHADERS

V-RAY - OBJECT PREPARED FOR V-RAY RENDERER (WITH TEXTURES AND SHADERS)

*. MAX 2010 - OR HIGHER

ELEMENT OF MODEL

MAX

3DS

FBX - SIMPLE OBJECT WITHOUT MATERIALS (WITH MAPPING AND TEXTURES INCLUDED)

*.MAX - V-RAY 1.5 - OR HIGHER - WITH TEXTURES AND SHADERS

V-RAY - OBJECT PREPARED FOR V-RAY RENDERER (WITH TEXTURES AND SHADERS)

*. MAX 2010 - OR HIGHER

TABLE-WARE VOL 1 MADE WITH 3DS MAX AND V-RAY

SLATE MATERIAL EDITOR

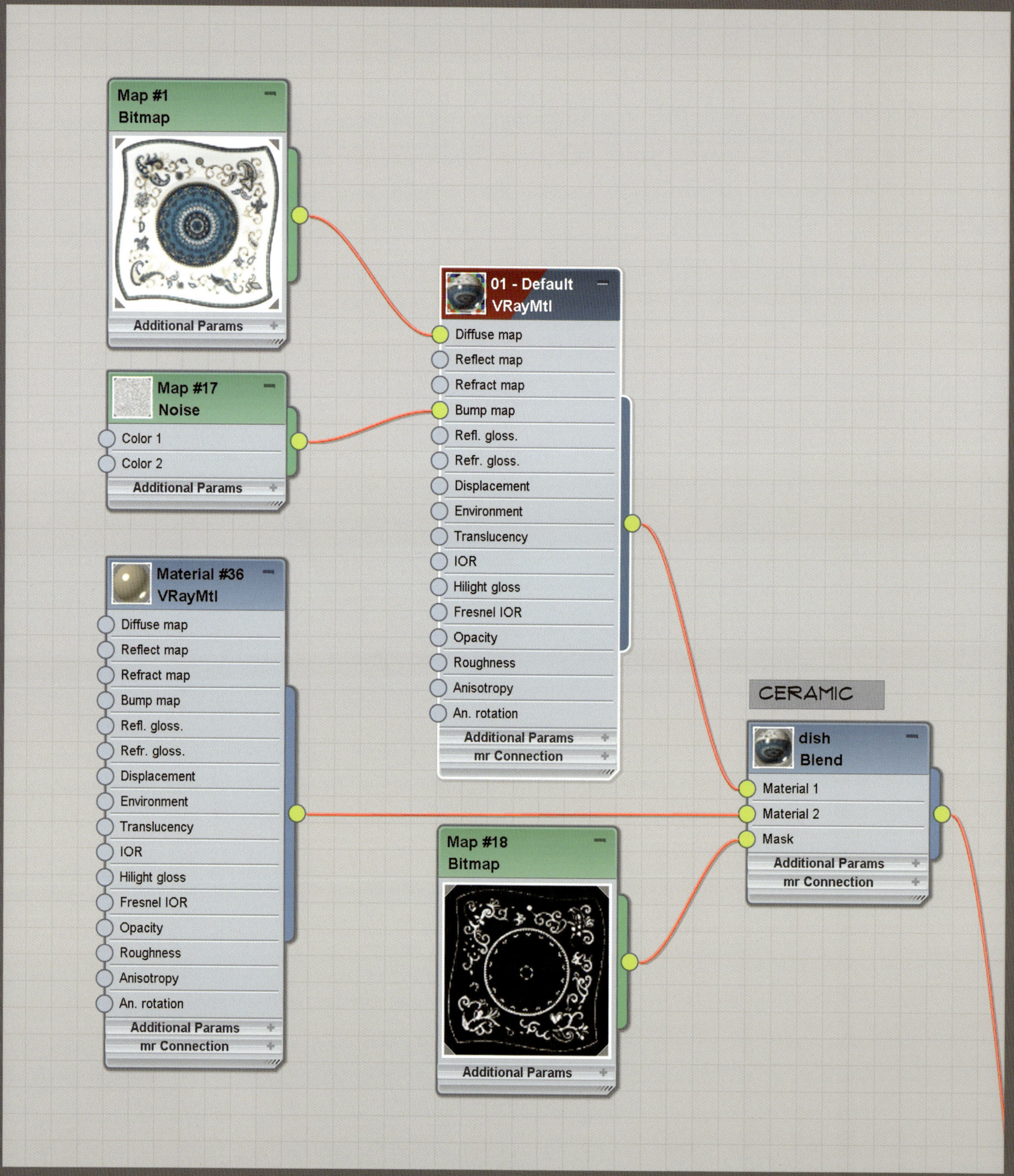

CERAMIC

TABLE-WARE VOL 1 MADE WITH 3DS MAX AND V-RAY

SLATE MATERIAL EDITOR

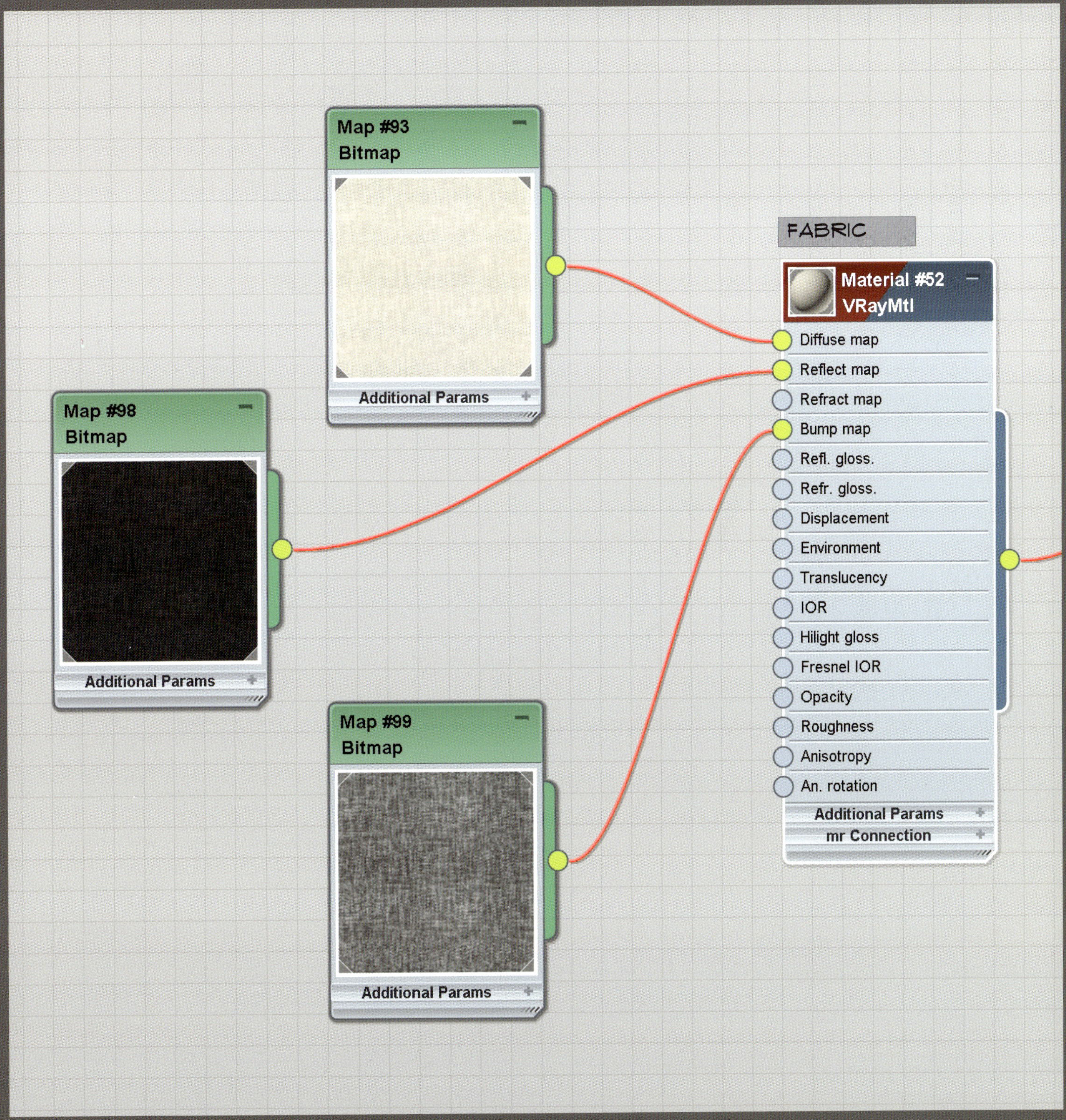

TABLE-WARE VOL 1 MADE WITH 3DS MAX AND V-RAY

ABOUT	SCRIPT

COLOR_CORRECT (THE METHOD WHICH USES THIS BOOK) COLORCORRECT_V3.4.98.12_X64_BETA

ColorCorrect interface is designed so that it represents the internal data flow. At the top you have the source color/texture settings. After that we have different sections each of which is applied to in order from top to bottom.

Color Swatch

if no source texture is specified, then this color will be used as the input color.

Source Map

This is the input texture slot. You can place any kind of texture maps here.

RGB Space (Pre-Process)

This section is used to pre-process the colors in RGB space.

Brightness

This value is added to all RGBA channels. Affects alpha too.

Contrast

This parameter scales the RGBA channel values around 50% intensity. So 0% contrast will produce 50% gray regardless the input color. Affects alpha too.

Un-multiply Alpha

If the source texture has alpha channel, RGB channel values are normally multiplied with the alpha channel value. For example if a white pixel has 50% alpha, it's stored as %50 gray. This provides faster compositing operations since RGB values are already multiplied with alpha. This is how 3ds max and most compositing software work and therefore you don't need to think about this normally.

However, if you want to change/replace the alpha channel, you should first find the original RGB values (before the pre-multiplication is done) and multiply it with the new alpha. This option does exactly that: un-multiplies the RGB color with the source alpha. This operation on a 50% gray pixel with 50% alpha will result in a white pixel. Also see the "Pre-Multiply New Alpha" option in RGBA Space Channel Mapping section

Clamp

When turned ON, RGBA channels are clamped by the specified min and max values. The normal range of the parameters is [0..1] but you can still specify values out of this range to clamp HDRI images. Affects alpha too.

Normalize

This option is available only when the Clamp is ON. When normalize is ON, clamped colors are scaled so that they map to [0..1] range again after the clamp operation (this provides you a control like the Photoshop's Levels setting). Affects alpha too.

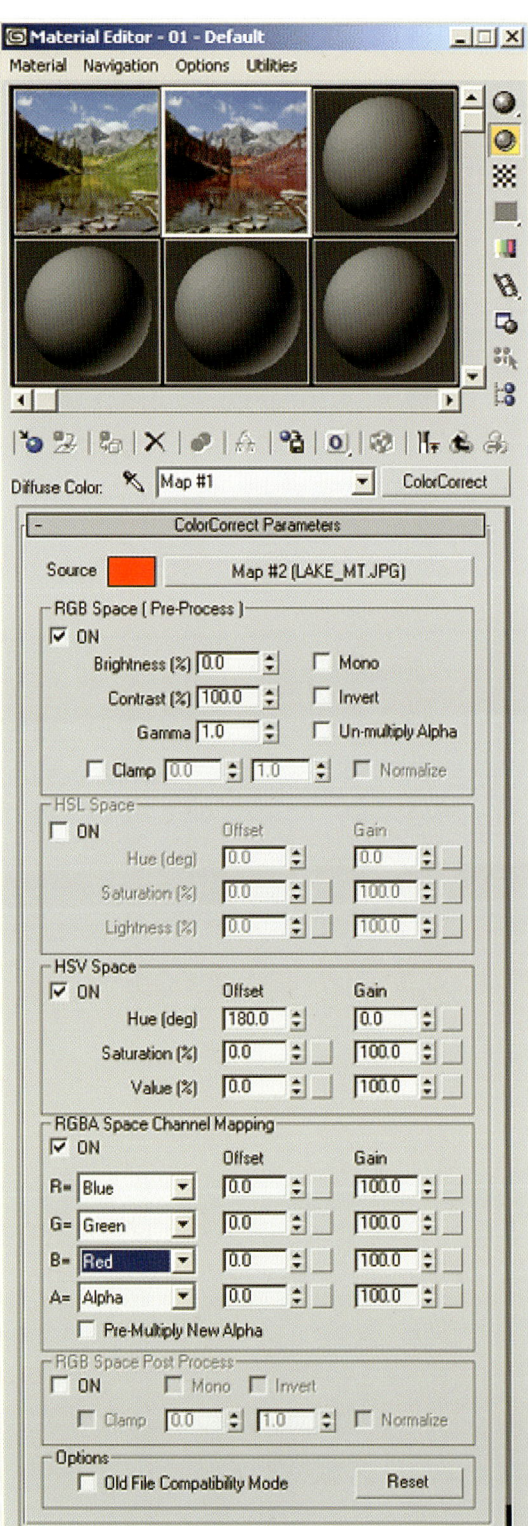

ABOUT SCRIPT

RENAME MAP   (THE METHOD WHICH USES THIS BOOK)   RENAME MAP.MSE(28.2KB)

This script shows a list of the current scene maepsoseu used in changing a name map script.

[Feature]
Edittext window at the bottom of the editing can change the file name (do not actually change maps maepsoseu file name is changed).

MatClean meditmeterial slot will be reset button (right-click and a little bit wrong)

Find Mat maepsoseuga button that shows real-meter was used.

Find Map button to map that shows maepsoseuga used.

And the presence of three modes: All, select object, missing map mode. (missing map also useful to free some space is used)

DEL button to select the map in real-Cleans meter (it's actually nothing to do maepsoseu files can be erased).

Change Map button maepsoseureul maepsoseuro can change the other.

Other different features, so please try.

*** 1.12 update ***
- The map to find the erased 'Del' Enhanced
- Finalrender material support - 'Find mat' Enhanced
- 'MatClean' right-click on the button Clear All Add meditmeterial

*** 1.01 update ***
- Rename Map of the features (improved search algorithm to map)
- Del's (spicy rainy season), strengthening the functions of
- ChangeMap enhancement of
- Refresh enhancements
- Explorer, select the features that add a file to open
- Map to selected 24 Added ability to display in real-meter (Find Map)
- Remove Features collect resources
- Open the selected file in Photoshop to remove a feature
- Max2008 more bug fixes, I have an error in.

CLEAN_MOTIONCLIP   (THIN INCREASED MEMORY CAPACITY AND PROBLEM SOLVING)   CLEAN_MOTIONCLIP.MS(91B)

It is another object by an external source and are not in order

Throw everything in the scene suddenly eats memory capacity grows lengthy cast save time solves the problem Animation and the reactor to an external source and eliminates the remaining.

How to open a file that is the problem

If the downloaded scripts work on max drag go ahead with the opening of the script to run.

3Ds MAX 2012 64bit version of test